SHATTERED INTUITION

BY

A. J. HAPKE

✲✲✲

Published by Hemingway publishers

Cover design by Hemingway publishers

ISBN: Printed in the United States

HEMINGWAY
PUBLISHERS

* * *

TABLE OF CONTENTS

James 1:12:17

***

PREFACE

Freedom has been mine for a while now. My modest home is in a small suburb south of the city. Almost six years ago, I purchased a small 1950s Ranch-style home. It was move-in ready and adorable. The house has 2 bedrooms, 1 bathroom, and a basement. I use the spare bedroom as a walk-in closet. I have everything I need here, and I'm settled.

The commute to work is mildly brutal, easily over an hour each way; during rush hour. I'm the Administrative Officer for a company on the southern edge of the city. I'm blessed that the office is located right along the interstate. Each day starts early; I get dolled up for work, grab breakfast, and head out. There's a sandwich service that brings their food truck into the office plaza. *They make bank every day at lunchtime.* Each evening is the same: sit in traffic, get home, change clothes, fix a quick supper, pet the cat, and take a long hot bath before bed. It's pretty routine.

After several years on the job, I'm decent at it and can anticipate most issues throughout the day. I also prepare everything the execs might need for a meeting, phone call, or client. The computers are old, reliable, and nothing needs to be upgraded. It's always the same. Occasionally, I get to solve a paper jam or toner issue. Those instances provide a little excitement in the day. It's a good, consistent job that pays well; most people would tell me not to change it.

Each weekend consists of cleaning the house, paying bills, and running errands. The lawn guys take care of the yard; I usually don't even see them. Can you tell I'm bored? Yeah, I'm VERY bored. I've been thinking long and hard about changing jobs. *Maybe I should look for something closer to the house? I could get a dog then....*

Does anybody still post good jobs in the newspaper? On a lark, I pick one up and have a look. And there is a job listed for the local rock quarry. They need an "Executive Assistant" in their office. The pay listed is good and comparable to the job I have. The commute is 20 minutes – *only 20 minutes* – on a two-lane highway, no less. My only consideration is how dirty the place must be and if all my good clothes will be either too dressy or get ruined on the job. I decide to apply, and receive an email with an interview date and time.

CHAPTER: 1

I'm taking a vacation day. That sure set the alarm bells off at the office. I haven't used any vacation time since I started working there. I'm sure not going to tell them I'm interviewing for another job.

For the interview, I've decided to wear my usual slacks, heels, and sweater set. It seems silly that I would need to wear a suit for a job interview at a rock quarry. I decided to pull all my hair up today. I don't know these folks and don't want to look too severe, but it's a warm spring day, and wearing my hair up might make me less fidgety. If there's anything I hate to do, it's interview for a job.

Now would be a good time to describe myself: I'm 36 years old, 5 ft 7in, and curvy. I'm neither tall nor necessarily wide, but there's some hippage there. I'm blond and have blue-green eyes. Makeup for me is just some mascara and lip gloss. My jewelry is rather plain: white gold with small diamonds, stud earrings, a necklace, and sometimes a ring on my right hand.

I grew up in a small country town in Southern Illinois called Cisne. Right out of college, I followed my boyfriend to Chicago, where he was going to be a high-powered lawyer. The relationship with him lasted far too long. I've remained single, by choice, since. My degree is in Business Management, and I have always worked in various types of secretarial positions.

It's time to go! Driving 20 minutes to a rock quarry is not something I ever saw myself doing.

###

This is not what I expected; the GPS says I'm here. There is no sign for the quarry. I turn into the drive. The office appears to be located at the bottom of the quarry, which is surrounded by a massive, tall, horseshoe-shaped cliff of white limestone. I drive down the gravel lot to the base of the horseshoe and arrive at an all-white metal building. There are no other vehicles or equipment here, just an arrangement of white buildings somewhat mimicking the shape of the rock wall behind the offices. I pull up in front of what I'm assuming is the main building. It's constructed with a low-slung porch that wraps around the front and right side of the main "shed". Several feet to the left of the porch is a large double door, similar to the farm sheds back home. It genuinely looks like there is no one here. There's just my little dark red sedan parked out front now. *Um... how did everyone get to work today?*

I step out of the car and notice that there is none of the noise I expected. Nor any dust.... *What in the world?* Oops, I accidentally parked right in front of the walk-in door. I didn't see the door until I got out of the car. This porch is very deep, and the door is set far back from the edge of the concrete, surrounded by benches on either side of the door. The front door is a deep cherry wood color with a gorgeous cut glass inset in the main panel. There are also large pots of shade-loving flowers on either side; they look like they have just been planted. I'm trying to walk slowly and take in all the details.

Job interviews always bring out my nerves. *Deep breath...* I don't know whether I should knock or just walk in. I guess I'll just grab the door handle and go inside.

###

I look up as I step inside the cool office. Greeting me is a beautiful Native American woman smiling a wide, bright-white smile. She has very long, straight, jet-black hair, and her makeup is totally on point, even with green eye shadow. She has on yellow gold understated jewelry, a simple chain necklace, and hoop earrings. The nameplate on her desk says Maia.

"I'm SO glad you're finally here!" she exclaims, stepping around the cherry wood reception desk and walking forward to greet me. She grabs ahold of me and hugs me tight like we are long-lost cousins. She's quite tall. Maia is dressed in dark slacks and a deep green, button-down silk shirt. *Whew, I wore the right thing.* I don't know what to do with my arms, so they just hang by my sides. *Who knows what my mouth is gonna spout in this situation?*

"My God, you're stunning." Is all I say, stammering. *Geez, now she thinks THAT, and I couldn't even say hello or something professional.* My cheeks turn red. I look up at the sound of a deep African American voice. "Calm down, Maia, she doesn't even know where she is. Let her take a breath." He is chuckling and smiling a broad, toothy, and also bright-white smile. *Does everyone here bleach their teeth?*

He steps around the desk and walks toward me. He is tall enough to be a pro-league basketball player. His skin is quite dark; his face is chiseled and clean-shaven. The man's hair is shoulder length but done up in braids with part of the braids pulled back. He is dressed well, with pleated khaki pants made from expensive material. *Who wears pleats anymore?* His shirt is a button-down blue and white checked shirt that is also well-tailored. He sticks out his hand. "I'm Terrell from the Sales Department. Welcome to the office. Don't mind Maia, we don't get to meet new people very often, and she is so excited to have someone new to talk to." "Th Th Thank you." *I'm stammering again.* "I'm delighted to be here." I'm smiling as wide as I can. It's no match for either Maia's or Terrell's smiles, but I want them to know that they have made me feel as comfortable as I'm able to be in this situation.

This office is clean as a whistle. Neither Maia nor Terrell have a hair out of place, and there are no wrinkles in their clothes. Even their shoes are impeccably clean and polished. *Yes, I always look at people's shoes.* There are 3 other desks beside the receptionist's desk. They don't have the high front side that Maia's desk has but are open with bow fronts and storage cabinets up high against the wall. I don't see anyone else there, and the computers aren't turned on. There are real plants, that appear well cared for, arranged almost everywhere. To the left, it looks like there's a group of salesmen's offices with windows facing the main office area. There is a row of high-set windows on the right side of the space that allows alot of natural light in. *The windows must be set above the porch roof.* All of my wondering takes me away for a moment.

I turn back to Maia and Terrell. "This is a really great space. My interview isn't for another few minutes; could you give me a tour?" Terrell bows out and steps into the first office on the left. Maia walks me through the spacious area to another row of sales offices at the back of the large room. She points out that the other desks are vacant. I don't see any salesmen in their offices besides Terrell. Everything is quite tidy; not a single piece of paper is out of place. In front of the rear offices, she stops in the walking aisle and points, "To the left are the restrooms, and the dining room is directly across from them. Off to the right, you can see the executive offices and the conference room. Do you need anything before I take you to the conference room? Water? Pit stop?" "No, I'm fine, thank you. Is everyone off today? Is today a holiday and I missed it?" I cannot control my mouth when I'm nervous. *This is always a problem. Get a grip, woman.*

Maia laughs and says, "No, it's time for spring vacations before the summer heat takes over all the tropical locations. Almost everyone took off for a holiday to Bermuda this week. Just a few of us stayed here to keep everything going." My eyebrows raise, and I whisper-squeak, "Ber-Bermuda? Like the island?" *Way to go, Miss Intelligence.* Maia is still smiling. "Yes, the company has several resort holdings here and there. They sponsor vacations for us a few times a year. It doesn't count towards vacation time, either." *Um...WOW.* "That's a wonderful benefit. I've never encountered a vacation package like that in an office position."

She guides me to the right and into the conference room. It's the back wall of the shed-style building, and there's a row of wide windows at chair rail height. Rising above the back of the building is the tall stone edifice. It looks like there's also a bubbling, fairly-wide stream that butts up against the base of the rock wall. "This is beautiful!" I exclaim. Bless Maia; she just keeps smiling at me. "It really is beautiful, and it's a great place to work. The folks here are quite welcoming. I'm looking forward to working with you." *I don't have the job yet....*

Just then, Terrell walks up behind Maia and asks, "Are you getting her settled in? We need to grab lunch before that 1 o'clock." Maia nods to him and turns back to me. "Have a seat wherever you like and grab a bottle of water. Nathan is going to want to visit with you."

###

I'm not one to sit down in a space that I'm unfamiliar with. I walk over to the windows so I can watch the water. To my right and nestled in the corner is a snack station; including a bowl of fruit and some cellophane-wrapped snacks. The beverage cooler appears to be well stocked with water and soda. I lean over to grab a bottle of water from the fridge. *This will give me something to do with my hands.* My leather resume folder is tucked under my left arm. I fumble the bottle of water. My back is to the windows, and I'm facing the door. I step forward to grab the falling bottle. The fridge door is still open and as I turn away, I trip over my own feet. I catch myself with a loud bang, both hands slamming onto the food table.

The folder clatters to the floor, papers scattering. Of course, right on cue, an older gentleman walks in with a big, bright white smile on his face. *What is with the teeth?*

He is somewhat portly with white hair, a mustache, and tortoiseshell glasses. The man is dressed in dark blue jeans and a button-down shirt. He's wearing a 70's-style, avocado-green pattern sweater vest over top. His clothing is obviously expensive and well-tailored. *Clothes last for decades when you take care of them.* His shoes are a mahogany-colored dress shoe style. He isn't much taller than me.

"You must be Eleanor! Welcome to our little office!" With a start, he realizes that my face is quite red and I'm still tangled in the refrigerator door, leaning on the table. I look down at the mess on the floor. I'm not even going to try and reach it. "Oh, my dear, let me help you!" He walks forward and grabs my right hand, taking the bottle of water out of it. He lifts my arm and helps me steady myself back upright. My face and neck are maroon. I look down, step away from the fridge, and close the door. He still has ahold of my arm. "Thank you for your help." Is all I can muster.

This man is still smiling at me. "I would like to shake your hand, sir. Thank you for having me here." *Lost the job now, you klutz.* He lets go of my arm and steps back a half step. "Welcome, young lady. Where would you like to sit?" This whole situation has my nerves in a spin cycle. "Sir, excuse me. Give me just a moment so I can grab these papers. Do you have a preference for my location at the table?" I bend down to hurry up, grab the papers, and stuff them back into the folder as quickly as I can. He doesn't reply and sits down at the end of the conference table closest to the food. He smiles and says, "I find that I don't like to be very far away from a snack." And chuckles to himself. I sit opposite him across the short end of the table, laying the messy folder in front of me. I smile as sweetly as I can, even with mortification still written on my face. "Are you Nathan?" I ask. "Oh no, dear, he is on his way back from a meeting. He asked this old man to stop in and visit with you so you aren't waiting in here alone. As you can see, we have a skeleton crew this week. Please, call me Mr. Louis." I reply, "Thank you, Mr. Louis. I'm Eleanor Stanton; please call me Ellie."

Mr. Louis stays with me for a half hour. He gets me laughing, and I start to feel quite comfortable. Then, an extremely good-looking man strides in like he is in a hurry. He is tan, quite tall, with broad shoulders, black hair, and eyes the deep blue of the ocean. *Is my mouth hanging open? He should be a model for that underwear company. Where is my brain? Obviously not here...* He has on black pinstripe slacks, a light blue shirt, and his navy tie is undone. A matching suit jacket is cradled in his left arm. "I'm Nathan. Welcome to our business." He gives me a wide, bright white, and straight-toothed smile. *Seriously, the teeth....* We shake hands. "I'm heading out." Mr. Louis interrupts and pats Nathan on the shoulder.

"I, I'm Ellie." And I blush again, looking up at the tall man. Nathan smiles wide. "How has your visit been so far, Ellie? We are so glad you've decided to visit with us today." And the conversation goes from there. We discuss the business and my role in the office for over an hour. Then my stomach grumbles as loud as I have heard it in months. *Oh dear…* I blush again.

"Ellie! I'm so sorry; I didn't even consider lunch. Let me get you something, it's past 1. You must be famished." Nathan appears quite concerned. "I'm okay, Nathan, thank you. I don't want to take up too much of your time. I really appreciate everyone's hospitality today already." And I blush *again*. "No, Ellie, let's go see what Cook came up with today." He looks serious. *Cook?* "If you insist, I hate to be such a bother." I am looking down at the floor at this point. *Shoot me now; I don't know if I can manage my table manners today, too.* Nathan leads me out of the conference room and back around to the restrooms and kitchen.

"Do you need a pit stop?" Nathan asks with concern in his voice. Yes, I do, and I take a couple of minutes to straighten my clothes and my face. I hear voices coming from the kitchen; they almost echo. *Must be a big kitchen.* I peer into the doorway to find a large dining room with several large round tables and, in the rear, a tall countertop absolutely laden with every kind of pie known to man. "Ellie! This is Cook. He takes care of all of us here, any meal or snack we need. We are so far out from everything in the city." Nathan smiles and moves his hand towards Cook.

Cook is a tall man, carrying some weight on his belly and his cheeks are rosy as if he has just eaten dessert. He is strawberry blonde, with green eyes that point to some Viking heritage. He's a big man, and if he wasn't smiling so kindly, I'd be concerned. Cook sticks out his hand, "Hi Ellie, what can I get for you today?" I smile wide because I am hungry, and it smells delicious in here. "I'm not sure. Are there any leftovers from lunch? Otherwise, something simple like a grilled cheese would be fine." *Where is this boldness coming from? I'm getting too comfortable here.*

Cook frowns and gives me a stern look. "You will NOT be eating greasy grilled cheese in MY kitchen!" He slaps the counter with the towel from his shoulder. *Great, now I've pissed off the Viking.* "Oh no, I'm sorry, Cook. I was just trying to be simple. I will have whatever Nathan is having." Nathan shrugs and states he has already eaten. *Now I've got to eat alone in front of him?* Cook sees my distressed face and chimes in. "How about some ravioli? It's quick and simple; I can get you back to work in no time." *I don't work here yet....* "Thank you so much, Cook." I sit with Nathan at the counter, and we visit about how nice the office is. Quick as a wink, it seems, Cook brings a big plate of ravioli, garlic bread, and a little bowl of salad with ranch dressing. *How did he know?*

Nathan stands to fill a glass with iced tea for me, and I dig in. I wolf the food down, not even thinking about manners. Honestly, it's the best Italian food I think I've ever had. *And prepared by a Viking chef... harrumph.* Nathan continues to make light talk about this and that, and then the conversation turns back around to the buildings and how they are all white metal.

"It was the cheapest!" He snickers a little bit and glances back to Cook, who's working at the sink behind the counter. Cook winks, and Nathan turns back to me. "Really, we didn't want to detract from the rock face. It's already blinding white limestone, and any other color would have messed up the whole look of the place. We don't need a fancy building.... On the outside." He gives me a little crooked smile and raises his eyes to mine.

"I'm finished with this Cook. I ate the whole thing!" And I grin back at him. "Thank you so much, truly." I turn to Cook and then to Nathan to impress upon them both. *I guess the meal helped bolster my self-confidence.* We turn to leave, and Nathan asks, "Would you like to tour the rest of the facility? It's a little empty this week, so the touring is easier." I nod, and he takes me through a side door between the restrooms and kitchen, which I didn't notice the first time. *It must be the warehouse.*

###

Nathan leads me into a huge open warehouse space. Very high up are massive steel trusses that hold the roof over the entire area. It's almost empty; there's a forklift and some pallets, a tall machine of some kind (*no clue what that is),* and the whole place echoes with every footstep *(dadgum heels)*. The space is also extremely clean. I turn back to him with a questioning look in my eyes.

He chuckles and points to his left, "See those double doors? You saw the outside of those when you pulled up. This is our load-out zone." I nod, and he continues, "If you look across the floor, there's another exit door that takes you right out onto the bank of the stream. There's a patio area there that is always shaded due to the shadow from the rock face." I'm confused as to why he is telling me this, but I say, "Ya'll *(ya'll?!)* keep a nice place up for your employees. This is quite a facility." He chuckles a little. "You haven't even seen the bagging floor!" We turn to the right and walk some distance back. The air temperature continues to drop.

We pass the tall thing I noticed earlier, and Nathan points, "That's the bagging machine. After the mineral is processed, it's bagged and palleted." He turns his hand to a specific area, and I squint to focus. "See the conveyor belt? The guys turn it upright when they aren't using the machine, which makes it easier to navigate around the warehouse with the forklift." I nod. *There's nothing to say when you're totally confused by machinery.*

Nathan gestures to me to continue to walk forward. "You didn't realize it just now but we crossed a covered roadway over the stream and are headed back into the limestone cavern behind the building." *Why, oh why, did I decide to wear heels today?* It was like the man read my mind. "I am so sorry; I must apologize again. You're not in great shoes for this trek, are you? We are almost to the staging area. I will get you some boots to wear with your hard hat." *That's going to look lovely with these slacks.* "Thank you, Nathan, that's quite thoughtful. I'm sorry I didn't come prepared for the tour. I should have thought of that." *How would I have known? I've never been to a rock quarry before.*

We continue to walk, myself rather noisily, as a kerchunk kerchunk sound echoes off the walls of the cavern with every single step. We stop at a very well-lit area that is set up almost like a large locker room. There are stalls with emergency gear: hard hats, headlamps, first aid kits, boots, gloves, coveralls, and even ATVs staged. All of this is ready to go deep into the quarry. Nathan looks at my feet, deciding that they aren't small, and hands me a pair of army green rubber knee boots. "Put those on. We can leave your other shoes here for now. Here's your hat!" and he sets the thing right onto my head. *Oh no, my hair!* Nathan cocks his head to the side and grins wide, almost goofy. I glance up at him and smile because I don't know what else to do here. He snags the keys off the wall hook and points to a shiny red machine that seats 2 people. "Hop in."

We drive down a long corridor carved out of rock and enter into the largest cave I've ever seen. It's just a huge open room, bigger than a football stadium. There are no columns left for support. There are lights everywhere, but not so many I would call the room well-lit, almost like lamps used in a living room. The entire area sparkles. *Shimmers?* My mouth drops open. All over the walls are clear-to-white crystals of some type. From the equipment that is sitting around, it appears the employees drill into the walls to extract even more. There are small tractors with loaders on the front that they must use to scoop up what has been mined. Again, everything is extremely clean. Zero dust. "What is it?" I blurt out; I'm just so awestruck by the sight.

"It's a natural compound used by nutrition companies for vitamins and health supplements," Nathan spouts in a flat tone, almost like it's rote memorization. "Um, ok", I reply. "Let's head to the transport area" and he directs me back to the ATV. I didn't initially realize it, but on the left side of the room is a tunnel heading out, and there is a train of more ATVs, each with a bed on the back and an additional wagon hitched behind it. *Where's the ventilation for all this? Why isn't there any dust? What about the fumes from the ATVs?* "There's enough room beside the train to follow through, and you can see the whole process from start to finish. Hands and feet inside the ride, please!" Nathan chuckles again. *Is he as nervous as I am?*

Turns out the ATV train leads back around to the bagging area. Nathan points to a squat machine sitting behind the bagger and conveyor belt. "This is the crusher/sorter. It sorts the crystals into the various particle sizes coming from the quarry. It also will crush product to a smaller size, if necessary. Then the crew uses the conveyor system to send the different size crystals to whichever packaging line it needs to go to. There's a bagging crew and a warehouse crew. Do you need a pit stop?" He asks. "That would be great, thank you," I say.

He leads me across the warehouse area to the door he pointed out earlier. But there is no immediate exit from the building. There's another set of restrooms and an additional kitchen/dining area almost identical to the one in the office. I can see out the windows of the dining room; there's the patio he was talking about earlier. *Wow.*

I finish in the restroom and realize that I still have these infernal boots on. *Oh no.* I step outside, and Nathan is waiting for me, holding my heels. "Looking for these?" I smile at him. "How did you get back to them so quickly?" "I kinda know my way around here." He gives me a crooked half-smile. "Let me take this," as he lifts the hard hat off my now thoroughly mussed hair. There's that goofy grin again. *Shit.*

"I've kept you far too long today. It's almost 5 pm! Let me show you to your car," as he starts walking swiftly across the warehouse space. I can't keep up in my heels. I'm not sure what I've done or said wrong, but I'm being dismissed. And right now. *Damn, this would have been a great place to work....*

By the time I pull away from the quarry *(more like a mine)*, I'm already in my head. *What did I do wrong? What did I say? Should I have refused lunch? I'm never going to hear back and never going to get to work here.*

I don't have too much time to contemplate what happened before I pull in the drive at home. Even the cat knows something's up.

***

CHAPTER 2:

Sleep eludes me. *My face is going to show it, too.* I walk to the bathroom to do my business and look in the mirror. *Why is my skin so smooth today? It must be the right day of my cycle, I guess. No bags under my eyes; that's cool.*

I finish my usual morning routine and drive the almost hour to work. I set to going through the motions of the day. One of my coworkers comes up to me. "How was your *VACATION* day? You never take a day off. Why are you so glum? Didn't you have a good time? Did you go home for the weekend?" I mumble, "I'm fine." And turn away from her.

The next 2 weeks are more of the same. I don't know why this regular routine upsets me so much. *I guess now I know what the other side looks like and I was looking so forward to a change.*

Another few weeks go by, and it's hot, mid-summer. I hurry to my car after work and start it so I can crank the air conditioning. Once it starts blowing cold air, I get inside and back out of my parking space. Suddenly, my cell phone rings. I don't recognize the number at all, but the voice I recognize immediately, "Ellie my dear! How are you?" "Mr. Louis! I'm so glad to hear from you!" I exclaim. I jam the car into park, halfway in and halfway out of the parking spot. My coworkers all turn their heads to see if anything's wrong. I wave them off and pull the car back in the spot. I point to the phone. A couple of people nod and go back to their own vehicles.

"Ellie, would you be able to come down here? We have a situation, and we need your help." Mr. Louis sounds like he's in a bit of a panic. "Sure thing, Mr. Louis; I would be happy to. It will take me about an hour and a half; I'm in the city." *Sorry, Mr. Mittens, you're going to have to wait for your supper.* Mr. Louis continues, "No problem, Ellie; I know you're going to be able to solve this right away. Thank you for being willing." And then he hangs up. *What in the world?*

I navigate the traffic and race *(ok, not race, can't do that with the traffic like this)* to the quarry. I don't worry about my hair, makeup, shoes, or how badly I have to pee. Maia greets me at the door, that amazing smile already in place. "Hey, girl! Welcome back! Mr. Louis is in the conference room." I walk straight back to the conference room, waving at Terrell as I stride by. Mr. Louis stands to greet me and smiles that wide white smile. "Thank you, Ellie! I truly owe you!" I tilt my head, "How can I help you, Mr. Louis?"

In walks Nathan. "Ellie! What are you doing here?" Mr. Louis grins and strides out of the conference room without saying a word. I throw up my hands, "Hi Nathan, do you know what Mr. Louis needs help with? He called and said he needed my help." Nathan just stands and looks at me, staring into my eyes. He sighs, "Yes, Ellie, we do need your help and badly. I'm afraid I can't allow you to leave if you agree to help. We haven't had any time to discuss this, and it hurts my heart to do this to you. I know you have a life outside of your current job, and if you agree to continue to help us, that will change."

"Nathan, I'm confused. I interviewed for a position in your office, not to join the family. As for a life outside of work, I live alone with a cat and have a routine I follow every day and every weekend. I own my home. I haven't traveled home since my parents died, so that's out, and most of my friends have their own families now. Would you still allow me to talk with them from time to time?" My eyebrows have to be at my hairline at this point, and my throat is parched. *Where is this overt confidence coming from?*

"Sit. Please. Let me explain." Nathan sits, and I follow suit. Cook rushes in and distracts us both, "Ellie, have you eaten? I made venison stew and homemade biscuits today. Nathan, would you enjoy another portion?" I smile at Cook, "Thank you, Cook, that would be awesome." At the same time, Nathan nods to Cook, indicating that he would like more stew. Cook rushes back to the kitchen.

I turn back to Nathan, "Nathan, what is going on? This is making me uncomfortable." I stammer at him with an intense look. "I'm going to need to go back home tonight; I have an animal to care for."

Nathan stands up to dim the lights in the conference room, and a large TV screen at the back of the room comes on. It's a video feed from the warehouse. There are wrapped pallets full of white bags, and those pallets are stacked to the rafters. The warehouse appears to be completely full. I see no workers, and the machinery couldn't move through the space if they wanted it to.

Nathan stares at my face. "I know I can trust you. I've known that since your interview. I'm sorry I haven't called and I've allowed things to get to this point. Maia has too many things on her plate, and I attempted to take on some of her scheduling work until we could locate someone to help her. She doesn't know that I wasn't able to get the scheduling done either." He swallows hard. "Our business is growing, and even though I'm the boss, I cannot do it all like I used to. Several of our warehouse employees have stayed in Bermuda all this time as they didn't have an assignment for when to return. The crystal harvesters returned and have been bang-on getting their work done." He looks sad and turns his face to me. "I haven't called you back because I felt like I had already told you too much during the tour, and you wouldn't want to work here. I haven't interviewed anyone else because you were the top choice from the beginning. Based on your resume, you've worked with the same scheduling software in the past and would be able to pick up where you left off, even though it's been updated. Dad saw me struggling and apparently decided to step in." He looks down, miserable. I lay my hand over his on the tabletop and smile. "Thank you for telling me; let's get to work."

Cook walks in with dinner, and it is absolutely delicious, as before. My table manners fail me again. *What is with this food?* Nathan smiles, "Let's eat and then get to work." He inhales his food as well. After we finish eating, it's like radar, and Cook shows up to grab the plates. Nathan stands and points out the door, "Take the desk to the left of Maia's. She is looking so forward to being your work neighbor."

###

The office area is dark. Maia and Terrell are gone. There is a single sales office still lit. Inside sits a blond man, impeccably dressed and groomed. He waves at Nathan as we walk by. "That's Caleb. He's working with Terrell to get product sold. He's been here several years now." I smile and walk past the front of Maia's receptionist desk and sit down at the beautiful cherry wood desk next to hers. This thing has to be almost 10 feet square, in a U shape with storage everywhere, and it's equipped with every type of computer, phone, and radio needed to manage the entire complex.

Nathan interrupts my thoughts. "It's already 8 pm. Why don't you go home for the night? We can start on this in the morning. I will contact your previous office and have your belongings transferred here. What time could you be here in the morning?" I figure quickly in my head, with the time savings on the commute, "I think about 7, if not a few minutes earlier." *I'm forgetting something....I've missed something here....* Nathan smiles that mega-watt smile, "That is perfect. I will see you in the morning." And he walks away, leaving me sitting in the dark of the still-unfamiliar space by myself. *Oh no, now I'm going to trigger the alarm or something.* But it doesn't go off when I open the front door. *Don't they lock this place up?* I turn back to try the door handle and it opens right away. There's no one else around. I shrug and get into my car. *Heck, I even left my keys in the car when I arrived earlier.*

I hear a faint engine. I turn and see a sleek, dark Jaguar, windows heavily tinted, drive past me. The tires are crunching the gravel as it leads the way up the drive to the two-lane highway. I drive the glorious commute home and think about how quickly things can turn on a dime. The same old routine and nosy co-workers are gone as of tomorrow.

Mr. Mittens greets me at the door. Normally, he just yawns at me from the couch, but I'm super late tonight and obviously in trouble. Mr. Mittens isn't vocal often, but tonight, he is telling me all about his dinner being delayed. "Rawr, meow, rawr meow" (*feed me, servant*). I sigh. *What did I miss earlier?* My brain is working overtime. I'm over-tired and running on adrenaline. I bend down to give Mr. Mittens his food. He snubs his nose at it and wanders away, tail high. *He really is pissed.* I go through my usual nighttime routine and get settled in bed. It's still nagging at me. *What am I forgetting?*

Mr. Mittens doesn't come to snuggle me in bed either. I sleep fitfully but definitely not well. *First day of the new job, and I'm gonna look like shit.* I look in the mirror, and just as before, the expected eye bags aren't there. My skin is almost glowing. It's not even the good week of the month this time. *Weird.* I finish getting ready, grab a granola bar since I want to arrive early, and head out to the "quarry." I've also forgotten that I was forgetting something.

CHAPTER 3:

I drive up to the office at 6:45 am. Nathan is standing at the door waiting for me. *How did he know I was running early?* I smile and walk in; he directs me to my desk, and all my belongings from my old job are already here. *What the...?*

Nathan clears his throat. "I was able to contact your old boss. He stated he noticed that you had been down in the dumps lately. He wishes you well and met me at the office to gather your things last night. He said to let you know that he hates to see you leave but understands that this is a great opportunity for you." *How in the...?* "Thank you, Nathan." I manage to squeak out.

I hit the button on the computer, and in a flash, the welcome screen is up. "Is there a password?" I ask, as there has been at every other job I've worked, but then I remember the front door being unlocked last night. Nathan grins and says, "No. Passwords aren't needed here. Everyone has access to everything. If we cannot trust them, why are they still working here?"

I think my mouth must be open because he turns my chair back around to the computer and leans over to point out the icon for the Volley scheduling software. *Maybe the programmer liked volleyball?* I double-click and get into the software. *Holy schnikes...* I ask, "Nathan? There are over 10,000 employees here. I had no idea this facility was this large."

That smile returns, "This scheduling is for all the facilities nationwide. We are the hub of everything here, but there are locations all over the country. The only scheduling we don't handle is for the international locations. Since the time zones are so different, we leave the scheduling up to the individual offices. You can access their information, but it's not your responsibility to schedule their employees." *Gulp*.

"Sir, I'm in over my head already; I've never worked with scheduling this many employees, departments, and locations simultaneously." I look up at him with tears brimming in my eyes. My face is turning red. *I'm SO thankful Maia isn't here to see me lose it.* Nathan leans on the edge of the desk, "You're going to do fine with it. Once you learn the rhythm of things here, you will get through the scheduling in a flash each day."

"EACH DAY?" I squeal loudly. I start to stand up and reach for my purse. "Nathan… I…" He puts his hand on my shoulder and gently guides me back to sitting. "You already know how to run the software, and that's half the battle. Please stay and help us out. No one here is going to hold it against you if you mess up a schedule and somebody works overtime or stays an extra day on vacation."

My eyes are like saucers. I cannot believe what I'm hearing. Not only did I not realize this was an international corporation, but the amount of details needing to be volleyed about just seems insane to keep track of. *Volleyed... ha! That's right. I forgot something last night. What did I forget? There it is... Shit that isn't a bit good.*

What would I do with Mr. Mittens? What about the rest of my life outside of work? And I'm already here today; how did I miss that?

Pasting an innocent look on my face, I turn to Nathan. "Nathan, why did you tell me last night that I wouldn't be able to leave here if I started to work for you?" His eyebrows go to his hairline, and if they could have gone above it, I think they would have. He tries to smile, but it fails. "Ellie, I think you're misremembering some of the things we discussed. You don't need to leave your home to work here. I was very stressed last night and side-lined by your visit. I'm sorry if I relayed incorrect information." *I wonder if he has his fingers crossed behind his back. That is not what I remember him saying. I was looking for a job when I found this one. I can find another. Just make it through this day. Anything I cannot stand is a liar.*

Cook walks in through the front door. "Nathan! I was looking for you. Could you help me decide on the menu for today's lunch and dinner?" He looks down at me, "Ellie?! You're back; I'm just so glad to see you. Have you been into the dining room for breakfast yet?"

Cook gives Nathan a stare-down. Nathan frowns and turns to me, "Ellie, I'm so sorry. I seem to forget myself when it comes to you and food. There's breakfast every morning, so you don't have to worry about grabbing something on the way to work each day. Would you care to join us in the dining room?" Cook is still staring Nathan down. He shrugs and looks back at me. "Ellie?" Reluctantly, I stand and follow them. That granola bar hasn't held since I've been so stressed. I might as well eat and feel better before I start trying to dig into all this.

I'm the last one into the room, and Cook immediately sets down a plate overflowing with food: eggs, toast, bacon, sausage, hash browns… there's even a piece of ham under the eggs. The eggs are cooked perfectly over easy, just how I like. It smells wonderful, and I look up at Cook. "I suppose I will end up eating all of this and in a flash, too, won't I?"

Cook chuckles and fires back, "You had better; I've been working on breakfast since early this morning." *What is with this food?* My mouth is already full, and with the next forkful already heading to my mouth, I spit out, "Cook, do I need to make the schedule for your hours too?" Cook looks from me to Nathan and back again. "Um…"

Nathan interjects, "Cook's hours are set where he wants them. You will see him on the list, but you don't have to schedule him because he is a one-man show around here. Excuse me, Ellie." Nathan turns to Cook and follows him into the kitchen area behind the main counter. They start glaring at each other, and Cook is gesturing back at Nathan while they talk in hushed voices; at least, I *think* their mouths are moving. *I've never seen anybody get that worked up about food before.* I just keep eating.

After several minutes of basically watching these two whisper-argue, I've finished my food. I take my plate up to the counter, interrupt them, thank Cook for the food, and head back to my desk. Sitting down, I wiggle the mouse to wake up the computer and take a look around to see what other software programs I might be familiar with. There are all the usual office-related programs, an inventory program, and an accounting program.

Then, I see an icon labeled "Syncope." A quick Google search yields a definition: "1. Temporary loss of consciousness due to a fall in blood pressure. OR 2. The omission of sounds or letters from within a word, for example, when *library* is pronounced." What an odd name for software. I google the software but nothing turns up. *Well... I pronounce all the letters in library when I say it...*

Just then, Nathan joins me back at my desk and I click on the icon for Volley again. "I've already set up your email for you," Nathan explains. "Oh! Thank you, I hadn't even thought of that." I reply. My brain has been swimming since yesterday. Nathan looks down at his watch and curses. "I've got to go, Ellie; I've got a meeting at 9 that I need to drive to." He says drive like it's a chore, and he doesn't do it all the time like the rest of us.

He interrupts my thoughts with, "Are you settled enough to start getting acquainted with the program again? You can flip through all the employee information and get familiar with their locations and departments. Maia should arrive anytime here and be able to help you get a handle on the usual daily schedule. Thank you again for your help with this." Before I can turn around and reply, he is gone; I don't even see him walking away. Just a moment later, that same dark Jaguar passes by the front of the office. It's a deep blue, but the windows are totally blacked out. *At least I know what the boss man drives now.*

Maia walks up behind me and startles me. I must have been still staring out the front window. *How embarrassing.* "Maia! You surprised me. How are you today? Where is your car?" She smiles with all those white teeth, "Oh, I just live around the corner; I came in the back door." I look at her kind of surprised, still, and manage to reply, "Oh, that's nice for a commute! Would you be able to help me with this Volley program? I'm trying to get up to speed to help with scheduling." *How can she live around the corner? The quarry is surrounded by acres of farmland.*

Maia looks off to the side in time to see Terrell striding in from the back. He comes right up to us and greets Maia with a dazzling smile. *I wonder if there's something going on there.*

"Morning, Terrell, how have you been?" I interrupt their little staring contest to see if he will turn away. "Hi, Ellie! I've never been better. How are you? What's new besides the fact you've joined our team?" He slings an arm around Maia's shoulders. *Uh-huh...*

"I'm good, Terrell, thank you for asking. I'm a little stressed with this." And I gesture to the computer. Maia chimes in and shrugs out of Terrell's arm, "You're gonna do fine with it, Ellie. All of us had a learning curve coming to work here." She turns back to Terrell and gives him a wink. Terrell turns and heads back to his office, and I turn back to the computer. Maia leans over my shoulder and starts pointing out the different locations, the number of employees at each one and which departments they work in. It appears that this location is the only one with harvesting and packaging employees.

There are warehousing employees at each place, though. *We must ship out from here to all over the world. I've not seen a semi-truck pull in here once. But it's not like I've been here very long...*

"Maia, where do you think I should start with the schedule to get folks in and empty the warehouse?" I ask, not knowing I just stirred a hornet's nest. "What do you mean empty the warehouse? It's empty; I just walked through there this morning." She looks at me, flabbergasted. *What in the world?*

"Maia, Nathan showed me on the security cameras last night that the warehouse was completely jammed full, and the warehouse staff hasn't been back from Bermuda to ship the product out. That's why he wanted me here early today, in order to get the scheduling back under control." I'm sure my face shows my confusion. Maia grabs my arm and says, "Come on, let's go look."

We walk back to the door between the restrooms and dining room, and Maia flings the door open. The warehouse is completely empty, except for the machinery. Not even a stray pallet. *How in the....* Maia turns red in the face and starts to say, "I'm gonna tan..." Flummoxed, I interrupt her and ask, "How could this warehouse have been emptied overnight, Maia? There's no one here."

And just then, Cook walks up behind us, "Morning, ladies! Anybody need a cup of coffee?" *How can you turn down Cook?*

So Maia and I stop, each grabbing a to-go cup and fill them from the umpteen pots and varieties of coffee, tea, and flavorings. I was able to create a chai tea latte from scratch, just the way I like it. *There are a lot of perks here... Ha! Perks...*

Maia and I settle at our desks. The original issue of shipping products has apparently been remedied. Now, I just have to get a handle on who goes where, when, and for how long at a time.

I decide to start working on this location's scheduling. Suddenly, a phone call rings in with a Texas area code. Maia puts the call on speaker, "Hi Jim! How's things in the Lonestar State?" Jim's reply isn't kind, and Maia takes the call off speaker, snatching up the handset as quickly as she can.

I overhear her side, though. "Jim, I don't know what happened. I had no idea things were backed up here until Ellie mentioned it this morning. What do you mean? THREE trucks pulled up there just now? No, I know your warehouse isn't that large. Go ahead and unload what you can. Let me call the shipping company and see if they will allow us to rent the trailers for a time. Thanks for being patient, Jim."

She hangs up and looks over at me. I realize I've been staring with my mouth open. "Ellie, it looks like Nathan shipped all this out last night, and it's just starting to arrive at the various locations. I might need your help securing some extra storage for them. We will probably hear from Wyoming next. These guys all drove overnight with the product."

I look up the town for Wyoming, a little place called Wheatland. I google it. *That's not much bigger than back home...* I call the various shipping companies in the local area there, and I'm able to secure 5 empty trailers to be delivered to the warehouse. I'm still so new that I don't even know how to find out how much product was shipped. Maia tells me to find out what can be done, so that's what I do. The various companies state they can deliver the extra trailers this morning before they start shipping out loads for other customers. *Whew.*

I look over at Maia between calls, "They are charging us a premium for all this." Maia shrugs, "It's ok; it's a necessary expense." I shrug. *Whaaaateverrrrr....*

I ask Maia what location will be next, and she thinks for a minute, "Oh crap, Ohio! Let me call them!" She dials their number and lets them know what's going on. She gives me the name of the town and I go back to calling the trucking and storage companies over in Ohio. This goes on for a couple more hours until we've arranged atleast some extra storage for each location. Only 1 shipping company out of the dozens I call aren't willing to help. *Guess they don't want our money...Our?* I look back at Maia. Her eyes are wild, and I think my hair is probably a mess.

Terrell walks in and asks if we are ready for a break. *I'm sure he overheard our conversations.*

Just then, the blonde salesman walks up and introduces himself. "Hi, Ellie. I'm Caleb. I'm in sales with Terrell. Thank you for all your hard work this morning. What a way to get baptized by fire!" He winks at Terrell, who, in return, grins back at Caleb. I shake

Caleb's hand and ask if he is ready for lunch, too. We all might as well go eat. "I'm famished," I say. *Confident much?*

Caleb leads us past the dining room and through the empty warehouse. We arrive at the second dining hall. It's set up a little differently than the office dining room, but only to take advantage of the view of the patio and stream outside. I walk into the open and bright space and see Cook standing up by the front counter. I wave and blurt, "Hi! Cook!"

Caleb stops me and corrects me, "That's Chef. He's Cook's brother. See how his hair is a bit darker?"

I walk up to Chef, "Hi Chef, I'm Ellie. I work in the office with these folks." Chef replies *(his voice deeper than Cook's)*, "Good to meet you, Ellie. Thank you for coming over here for lunch. It's been kinda lonely today."

Chef looks over at Terrell, "Do you know when the warehouse staff are returning?" Terrell looks at me, "Ellie? Do you know?" I blush red and reply to Chef instead of Terrell, "Chef, I'm just getting started on the scheduling. I hope to have everyone back and working within the next few days."

I glare at Terrell, as he knows what we have been doing all morning - covering Nathan's ass. Terrell just smiles that big-white-teeth smile back at me. Chef interjects, "What would ya'll like for lunch today? I haven't made anything specific, as the harvesting crew really only wants breakfast and dinner."

My first thought is Cook. *Will he be upset we came over here to eat?* And just like he heard me, Cook walks in. "Hey brother, what ya fixin' today? Looks like my crew came over here, so I thought I would too." *Whew.* Between all of us, we decide on steak salads with bleu cheese.

Cook sits with us as Chef starts lunch. I ask him, "Looks like both of you keep to your own kitchens then?" Cook chuckles a reply, "Oh yes, do not mess with a man's kitchen." He turns and continues his conversation with Terrell.

Breezing in through the side door is a short but beautiful woman with shoulder-length brown hair. She looks pregnant. She is holding a black cord in her hand above her head. She makes a beeline for Caleb, "Darling! Darling! You forgot the cable to your laptop this morning!"

Caleb stands and wraps his arms around her. "Amelia, baby! Thank you so much!" He turns to me, "Ellie, Amelia, Amelia, Ellie." And then he walks away to talk to Chef. "Hi, Amelia," I say, shaking her hand. "Would you like to join us for lunch? It looks like Caleb just went to order you a salad." Amelia smiles *(there are those teeth again)* and takes a seat. Amelia and Maia go right into a conversation about our morning.

Terrell turns to me, "How are you feeling? It's been a crazy first day for you." I agree that it has been nuts, but I'm looking forward to getting a handle on things. *Unless Nathan decides to lie to me again...*

Lunch is fantastic, and we adjourn to our respective workspaces. Amelia kisses Caleb on the cheek and heads back out the side door. As we are walking, I ask Maia, "How have you been for the past few weeks since I wasn't here? Anything new?" Her face falls. *Crap, I've said the wrong thing.* My face shows concern, and she interrupts my thoughts.

"Oh no, nothing horrible. I'm just weary. I needed that break to Bermuda to regroup, and this fiasco today sure reinforced it." She gestures to the mess of papers strewn across both our desks. I ask, "Do you think if I pick things up fast enough, you might be able to take a few days off?" She smiles a little half-smile. "That would be lovely. We will see how things go. There are still 2 more desks to fill, as you can see. A couple of the sales guys have been traveling and trying to recruit to fill the offices in the back. We are just so short-staffed." I gasp, and I think it takes her by surprise. I say, "You mean the office is meant to be full of folks?! I thought they were just extra spots for growth." Maia replies, "Oh yes, each desk used to have someone working at it. And the sales offices used to be full; save a couple for meetings and spare workspace." *Well, I can't quit now....* I stammer, "Maia… If I had any friends…" She interrupts me, "No, dear, there are only specific people that Mr. Louis and Nathan will entertain hiring. You had very specific qualifications that made you eligible for the position." "Oh," I whisper, flummoxed.

After we return to our desks, I work in the Volley program and Maia fields the calls coming in from the various locations complaining about the extra stock that Nathan shipped out overnight. By the end of the day, I'm hungry again and really tired. Cooking dinner isn't even remotely appealing and I decide it's a good pizza night.

I grab my purse and wave to Maia as I start for the door. She asks the person on the phone to hold for a moment and asks, "Where are you going?" "Maia, I live 20 minutes away. I have to get home to Mr. Mittens and get some supper." Maia says, "Didn't Nathan tell you to eat here in the evenings? There are usually some folks gathering and either playing yard games or board games or catching up on the sports and news we miss throughout the day. You should stay tonight. I bet Cook would even fix that pizza you're wanting." *How does she know?* I wave her off, "Thank you for the invite, Maia; another night might work better for me. Have a great evening!" I walk out the door.

✻✻✻

CHAPTER 4:

W*HERE IS MY CAR?!* I didn't even notice it was missing earlier. Now that I need it, it's not here. Roaring around the side of the building and skidding into the parking spot is my car. It's very shiny and quite clean. It looks even more red than it did this morning. *What in the world?* A kid, no older than 19, hops out and literally tosses me the keys.

"Nice ride, ma'am." He strides away. I look all over around the quarry and up at the rock face. I see no one, nor do I hear anything going on. Just the rush of the stream rolling over the rocks. I get in the car, it's still running, but it's quieter somehow. EVERYTHING inside it is clean and shining. There is not a speck of dust; it's immaculate. I had no idea anybody would take my car and detail it while I worked. How many perks do they offer? We hadn't even talked about health insurance; since I've been healthy overall, I didn't worry about it.

I stop at the top of the driveway and call in my pizza order. *The local chain gas stations really do have the best pizza.* I pull into the station and stop at an open pump. First, I attempt to top off the tank, but it appears that the kid has also filled my gas tank. *This is insane. They can't make any money doing all this for their employees. I can't be the only one they are doing this for unless they are just trying to get me to stay....* I walk in and pay for the pizza, then buckle the seatbelt over it on the passenger seat. My stomach roars at the smell of the pizza all the way home.

Since I'm on time tonight, Mr. Mittens gives me the usual stretch and yawn from the couch when I walk in. *Oh good, I'm forgiven.* The pizza is good, but it sure doesn't compare to the food I've been eating at work. I don't know what Chef and Cook do to their food, but it's simply phenomenal. *If I keep eating like that, though, I'm not going to fit into my clothes.* The usual nightly routine commences, but when I step out of the bathtub, I get a glance of myself in the mirror. I've had some little dimples of cellulite starting on my thighs for years now. They are gone. *GONE!* Who knew I would get upset about missing cellulite?! My face is still smooth, like it was this morning. The blue-green of my eyes seems brighter somehow. This is just strange. *All I did was change jobs, not start a health kick...*

Morning rolls around, and I must have slept well because I don't remember falling asleep. The bedside lamp is still on when the alarm goes off. I am assuming I'm supposed to be at work by 8 am, as Nathan and I didn't get to discuss it. *I wonder where he was the rest of the day while we covered his rear?* I glance at the box of granola bars on my way out the door. I'm looking so forward to Cook's breakfast again. *I better get there a few minutes early so I can eat.*

I walk in the door at about 7:45, and I'm the first one here. *Crap, I don't know how to turn on the lights.* I wander around to the back and find Cook working away in the kitchen. Mr. Louis and Nathan are hunched over some paperwork at a corner table. Mr. Louis looks up and exclaims, “Ellie, my dear! How are you today?” *He's always so jovial.* “Hi, Mr. Louis, I’m good. I thought Cook might get breakfast for me. Would you happen to know how to turn on the lights in the office? It appears I’m the first one here today.” Nathan looks up, squints at me, and turns his head back to the paperwork. *That's a fine how do you do...*

Mr. Louis glares at Nathan, stands and walks over to me, putting an arm around my shoulders. “Cook! Hey Cook! You got another one!” Cook sets the pan off to the side burner and walks forward to get my order but says, “Same as yesterday?” I laugh, “Yeah, that sounds really good, Cook. Thank you.” He smiles and turns back to the stove. Mr. Louis still has his arm around my shoulder and guides me out of the dining room, right past the scowling Nathan. “Let me show you where all the controls for the office are.”

We walk back out into the office walkway, and he takes me to a narrow door between Terrell’s office and the second one in the row. I hadn’t noticed it before; it’s the same color and pattern as the wall. You can barely see the narrowest crack around the outside of it. It’s more of a panel than a door. Mr. Louis presses the panel in, and it pops open to reveal the server tower, all the electrical controls, and the security system panel. To the right of the black stand with all the equipment on it is a bank of switches. They are labeled, thank Heavens. Mr. Louis explains, “There are stations like this throughout the facility. We’ve hidden them so they aren’t as noticeable; it detracts from the aesthetic. The panel is also insulated to keep any noise from the machinery out of the office space. You will usually find the lighting and temperature controls within these panels." I nod in response. "Thank you so much, Mr. Louis; I will get to work." I reach in to flip switches as he walks away. I hear a crash in the dining room and see Nathan stomping towards his office. I'm afraid to look but walk cautiously back to the dining room to see if there's any help I can offer. Mr. Louis is red-faced, and Cook is bent over to pick up the completely upended table. The food and dishes are scattered everywhere. The chairs are pushed far away from the table. I bend down to start cleaning up broken dishes and glass. I find the pieces of flatware and gather those for the dishwasher. "My, oh my…" Mr. Louis mutters. I'm sure he is embarrassed, so I don't look at him. Cook runs to fetch a trash can and dish tub as we are working on the mess. Mr. Louis steps out of the room.

Just then, a short, pale, quiet-looking woman comes in and exclaims loudly, "What the hell, Cook?!" *So much for quiet...* "Nova, I didn't do this; I'm just cleaning it up." Cook rolls his eyes. He looks at me, "Ellie, meet Nova; she keeps everything clean around here. Don't get on her bad side; she will never let you forget it." I stand up and nod to Nova, "I'm sorry, Nova, could I shake your hand another time? Mine are pretty messy at the moment. Glad to meet you." She immediately bears down on me, "Did you do this? Why would you make such a mess?" I stand back quickly, "No, ma'am. I was just trying to help clean it up. I don't know how it happened, but I heard a noise and came to help." She glares at me like I must have lied to her. She leaves and almost immediately returns with a combination vacuum/mop and starts working on the smeared eggs behind us. I motion to Cook to flip the table back over, and he does it single-handedly like it isn't made of oak. I stare at him, and he shrugs. "Viking… remember?" *Riiiight.*

Cook smiles, "Go back to your desk, Nova, and I can get the rest of the pieces. Thank you for your help. You didn't have to." I shrug back at him, "I am happy to help. Sorry, this happened and ruined your morning."

###

Nathan had to be the culprit of the stunt, and now he's nowhere to be found. I walk down the hall to what I assume is Mr. Louis' office at the end of the row. It's large but still cozy, with deep brown leather furniture and a massive antique desk. Behind the desk is part of the same bank of windows that line the conference room. *What a beautiful view to see while you're working.*

I knock very lightly on the door. Mr. Louis is sitting at his desk, but his head is bowed down, and I hope he isn't sleeping. He startles at the knock, "Oh, Ellie, I'm sorry, I didn't see you there. Come in, come in. Shut the door, please." I walk in, quietly shut the door, and he directs me to a leather chair in front of the fireplace on the back wall. I sit and wait for him to join me. He remains standing, and I feel like I'm going to be interrogated. "Ellie, I have to apologize for Nathan; he is under quite a bit of stress." I interrupt the man, "Yes, he said as much the other evening when I was here. Is there anything else the rest of us can do to help him get through this?" *As long as he stops lying to me…*

Mr. Louis sighs, "Actually, Ellie, you're part of the problem. I don't think he really wants me to relay this information, but I must. Nathan has all of the normal stress of his regular position here. In addition, Nathan has taken on much of his brother's duties since he left. He really doesn't know how to say no. And he was attracted to you from the beginning. That is a hard pill to swallow for us. We are used to being alone, and adding a female to the team that we are attracted to just makes it that much more difficult." *THAT explains the attitude and the crooked grin.* He continues, "Nathan doesn't know it yet, but I've called his brother Malcolm back home. Malcolm left a little over a year ago due to heartbreak, and it's time for him to come home. Malcolm should arrive before the end of the week."

"Mr. Louis," I begin, and he cuts me off. "Ellie, you can't help who you are any more than how you look. You're perfect for the position here. Nathan hasn't followed protocol as he is a bit tender-hearted towards you. I believe you witnessed him discussing this with Cook?" *So that was what their whisper fight was about the other day.*

Mr. Louis continues, "I believe part of the issue Nathan is having is that you don't seem to return his attraction to you. We don't frown on workplace romance here. We have a special situation, and getting folks together seems to really help with long-term morale. In fact, several of the couples that have started a romance here have gotten married and started families. You met Amelia, correct?" I'm flabbergasted, and my mind is reeling. Mr.

Louis is staring me down, and I better answer him. "Oh yes, she seems very nice." He continues, "She is married to Caleb, she used to work in the office with Maia but decided to stay home now that they've started their family."

I clear my throat. *This conversation has to change and fast.* "Mr. Louis, I..." But he cuts me off again. "I've made you uncomfortable, and I'm sorry for that. You need to know that I keep tabs on pretty much everything that goes on around here." I try again, "Mr. Louis, I..." He smiles, "Ellie. I apologize. What were you trying to ask?" I try for a third time, "When you say that Nathan hasn't followed protocol with me, what rules hasn't he followed? I don't want to be doing the wrong thing."

At that moment, Nathan bursts in the door, murder written on his face. He looks directly at me, "Ellie, please leave." *Well, buddy, you don't have to tell me twice.*

***

CHAPTER 5:

I go back to my desk and start up the Volley program. Maia steps out of Terrell's office; her lipstick is smudged. I smile and hand her my compact so she can fix it. *Alrighty then....* "How are you this morning, Maia?" I ask with wide eyes. She giggles and turns to take the phones off the night message.

Cook runs in with my forgotten breakfast. "You're so awesome, Cook, thank you." He waits and visits with Maia while I inhale the food. "That was amazing as usual, Cook!" I grin at him; he takes my dishes, nods to Maia, and heads back to the kitchen. *What would we all do without him? We?*

By the end of the day, I have managed to call back about half of the warehouse crew. After discussing preferred work times/days with each one, I have ironed out the next 2 weeks of scheduling. There isn't going to be a back log in the warehouse now! *Yay!*

###

It's 5 pm, I'm worn out. Today has been a roller coaster of emotions. I pick up my purse to head home again. Cook walks into the office toting a stainless steel box about the size of 2 shoe boxes. "What's this, Cook?" I ask and smile up at him. "Maia spilled the beans about your pizza last night. I don't want you eating that crap, so I'm sending supper home with you." He grins a crooked grin. "I've wrapped everything separately. If you plug this into the power outlet in your car, your dinner will stay hot until you can get home to eat it." *Like it's so very far...*

Grinning ear to ear, I look into Cook's caring eyes, "This is incredibly sweet, Cook; it smells divine. I'm looking forward to eating it already." He smiles back and turns to leave the office. I pick up my purse and the box and head to my car. Maia walks up behind me and shuts the door since my hands are full. "Have a great night!" She yells back at me.

###

When I pull up to the house, there is a deep blue Jaguar sitting in front, parked at the curb. I look up and see the front curtains move on both the neighbors' houses. *Great, now I gotta listen to that on Sunday when they stop to talk.* I pretend that I don't see Nathan walking up to me. I step out of the car and lean into the passenger side to grab the food box. Nathan asks, "Could I carry that for you?" I look up at him, "Why are you here? I don't want to be rude, but this is my home, my own time, and I'm not at the office." His face falls, and I cannot bear it. I sigh heavily. "Nathan, maybe it's good you're here so we can clear the air on some things." He perks up again, and I hand him the food box. I unlock the front door, allowing Nathan in first. Mr. Mittens is so freaked at seeing another human being that he takes off as fast as his little legs will carry that squishy belly. *That's the last of him I will see until the morning*. I roll my eyes.

Nathan walks right into the kitchen to set down the box. He starts calling, "Mr. Mittens, here, kitty kitty. I've got your supper.

Mr. Mittens!" *Oh, for Pete's sake.* I'm laughing at this point because this tall grown man is in my house, calling "here kitty kitty" for a spoiled fat cat with an attitude that I'm certain will not give him the time of day. Nathan turns to me. "It's great to hear you laugh. The past couple of days have been stressful for us both." He walks right up to me, into my personal space. I place my hand on his chest, "This won't work. I'm sorry." Nathan pulls out a chair and plops into it. *I've never seen the CEO of a company lose his shit as often as I have today. What is the deal?* "Ellie, can I level with you? It might take me some time to explain." I look him up and down. He looks absolutely exhausted. "That would be great, Nathan; let me get this supper plated up. You know Cook always fixes too much; I know there's going to be enough for us both." He nods and goes right to the cabinet with the plates and the drawer with the silverware. *What the....*

Cook sent TWO huge T-bone steaks, 2 baked potatoes, roast asparagus, and 2 salads. He even included the butter and sour cream for the potatoes, 3 miniature bottles of steak sauce, AND 3 different salad dressings packaged in tiny containers. *Clearly, this visit was planned.* Nathan set the table in the time it took me to marvel at the food, which was held at the perfect temperature inside the fancy box. We visit about light subjects over dinner. After that, Nathan stands up, walks to the fridge, and pulls out a strawberry-rhubarb pie with a lattice top. *How in the....* Now I know I didn't transport the thing, and how did he know it was my favorite? *And how did it get into my house?!* I'll leave the discussion of those questions for a different moment. Nathan turns around and asks, "Do you have any ice cream? I completely forgot about ice cream." I motion to the bottom freezer on the fridge. "There's some French Vanilla in there, but I'm sure it won't be as good as Cook's ice cream has to be."

I'm already stuffed to the gills; how will I fit pie and ice cream in there, too? *Who cares? I'm gonna eat it.* I don't know where this appetite is coming from but my pants don't feel any tighter. If anything, they have been a bit loose the past day or so. I'm lost in thought and Nathan walks up behind me. "Ellie…" "OH!" I screech, and the ice cream dipper goes flying out of my hand. Nathan catches it. CATCHES IT. As in, he traveled across the room in the amount of time before the thing hit the floor in the doorway to the living room. I'm standing with the ice cream container in one hand and my mouth hanging open. Nathan holds up the scoop and grins, "I guess the cat is out of the bag."

I'm sure my face is red. I bear down on this tall, handsome man in my kitchen and start beating my fists on his chest. "What in the actual FUCK, Nathan!" He interjects into my rant, "WHOA! Hold up. Take a breath. I told you we needed to talk." I start to cry. If there is anything I hate about myself, it's that I cry when I'm angry. I'm so mad in this moment that I could bite ten-penny nails in half. *Really girl? You don't even know this man and after a couple days of working together, you're this worked up? This is not sustainable.* The ice cream is still sitting on the counter. Nathan calmly walks over, puts the lid on the container and sets it back in the fridge. He goes to the correct drawer, pulls out the plastic wrap, and wraps the pie and both plates. *My blood pressure reading? NASA.*

I grit my teeth. "Nathan…." I'm growling at this point. He pulls out my chair at the table and says, "Sit, Ellie, please." I glare but sit. Smart man, he sits across the table from me this time instead of the chair next to mine. "I would like to start from the beginning when we first advertised for your position." I take a sip of water and nod; talking wouldn't be prudent right now. My ears are open, however, and I'm hoping he clears up some things.

He begins, "We have had a few folks leave the office recently and needed to fill at least one spot right away. As you know, we take our time adding employees and search for the perfect candidate. You're exactly right for our office. I completed the background checks and work experience review on you before you even showed up to interview. I didn't need to talk to you because everything already proved that you are fully capable and would be a huge help. What I didn't realize was that you were so attractive. I have other words for it, but they aren't appropriate in this situation. Because I was so attracted to you, I made dumb decisions, even on the day of your interview. Folks don't get a tour of the facility upon their first visit. They are vetted so thoroughly and are already informed of our culture and housing situation prior to joining the team. You were so professional that day; I knew I was being an idiot." He gulps some air and decides to take a drink. I say nothing and attempt to keep my face passive.

"When Dad called you in the other evening, I was already at my breaking point. I had placed way too many burdens upon myself and he was attempting to help. He could only call *you*. There wasn't anyone else we had investigated yet. I didn't want you to join our team because of my attraction to you. I was afraid it would get in the way of working together, and it obviously has already. And I have to apologize profusely for lying to you the morning you started work. I know I said that you would have to live at the quarry instead of your home. I didn't want to freak you out, but I knew I screwed up the instant I tried to smooth it over." I nod, and Nathan continues, "Then the shipping debacle that morning… I've been so overwhelmed and trying to solve problems like I did a decade ago, doing it all by myself and forgetting steps. I know better. Thank you for helping Maia sort out the issues and getting everything back on track. It's my understanding that the warehouse crew is on their way back in?" I nod again; I don't trust my voice yet.

"Let me fill you in on a few things that you haven't been privy to yet. All the folks at the office have been wonderful in helping me sort this out for myself and not telling you things that I should tell you myself. I'm sure you've noticed everyone's teeth?" I am stunned, "Yes, I have. It's crazy, everyone has perfectly straight and perfectly white teeth." Nathan chuckles, "If you continue to work for us and eat meals with us, your teeth will end up like that, too. I'm sure you've noticed your body changing?" My eyes are wide, "Yes, I have noticed little things changing." He cuts me off, "That is the nutritional supplement that we harvest. It contains all the minerals needed for a body to thrive. Chef and Cook add it to all our meals. It's even in our coffee. If you have any old bone breaks, bruises, moles, scars, they will all heal."

I sit back in my chair, stunned. "Why isn't everybody..." Nathan cuts me off again. "We are allowed to consume it at levels not considered "safe" by the government. Since we have it right there and consume it right there, they have no say in the matter. The outside locations that we ship to are only allowed to sell a certain amount to nutrition companies as a supplement. Because they also eat meals at their locations, they are able to consume the mineral at levels not normally allowed." I gulp. I have no idea how to take this. Not only have I been eating a supplement I didn't agree to, but I didn't even have a choice in it. *Shows how much we blindly trust food as a society, eh?* "Nathan, I..." He cuts me off again. "Ellie, I know you're trustworthy, or we wouldn't be having this conversation. Over time you will notice an increase in your stamina, your ability to move and move very quickly. You will begin to opt to run instead of walk and choose foot transport over driving." *That's where it all ties in....* "Taking this supplement guarantees a very long life. Do you have any idea how old Dad is?" I stammer, "maybe 65?" Nathan laughs, a big booming laugh. "Oh, he will love that. He is 142." *One hundred and forty-two years?!* My brain shuts down. This is too much. I'm suddenly very tired. "Nathan... I..."

He cuts me off again, "Ellie, we really need to continue this conversation. Dad discovered the pocket of mineral decades ago. I won't go into the chemical analysis of it, but it's many hundreds, if not thousands of years old. Dad was farming the land surrounding the quarry and when he stopped to drink from the creek, he noticed that the water was incredibly fresh and clear in this one spot. Once he investigated, he found a white powder leaching into the side of the bank where the stream had cut through. The mineral tasted sweet, and after sending off a sample for contaminant testing and it was found to be safe; he started harvesting it with buckets and shovels. Then, he started marketing the mineral to nutrition companies for "health benefits." *Where is his mom?*

I'm overwhelmed. "Nathan, I have to stop here. I'm overloaded and exhausted. Could we continue this conversation tomorrow?" He looks at me gravely, "Yes, Ellie, we can stop. But only if you promise to show up to work tomorrow. I don't want to have scared you off." I agree, and he bids me goodnight and leaves. Just then, Mr. Mittens steps out from behind the couch and rubs up against my leg. There's a scrap of steak left on my plate. I set it down for him. *I wonder if it works on pets, too?*

CHAPTER 6:

I walk into the office at 7:59, not my normal schedule. My stomach is ravenous. But I just got here, so I set my purse down to get to work.

I hear a deep voice, similar to Nathan's but with a bit more of a timber to it. I turn around and lock eyes with **THE** most handsome man I have ever seen. Dark brown, lightly-tousled hair and hazel eyes, tan, very tall, and those broad shoulders that make any woman weak in the knees. He is wearing glasses, dark-tailored pants, and a deep maroon button-down shirt. His black cowboy boots are scuffed. *What in the…?* I don't get a word in edgewise before Terrell comes bounding out of his office. "MALCOLM, MY MAN!" They do their bro-style handshake and hug tightly. *This must be the brother.*

I turn back around and listen to Caleb and Maia greet Malcolm. I'm all fidgety. He's like a magnet. Finally, Maia brings Malcolm over to my desk and introduces me. I take off my computer glasses and set them down as I stand to shake his hand. *Did his skin just mold to mine?* "Great to meet you, Malcolm; I'm Ellie, the schedule girl." He laughs, "Schedule girl? We have a schedule girl, now? Where did you come from?" *Okay, rude.*

Maia sees my face and interjects, "Malcolm, we needed Ellie to help take over the scheduling as it's gotten so large with all the new locations. She's doing a fantastic job for us and even steps in to help with other projects as needed." Malcolm shrugs and raises his hand in a half-wave, "I'll see you around, Ellie." *What the…?* Maia rolls her eyes and walks him around to the last sales office to settle him in. "This is where you're going to be for now, Malcolm. Your Dad said he will add on more executive offices by next year." I hear Malcolm grunt, "That's bull…" and turn back to my work.

Cook shows up at my desk, I didn't even hear him coming. "I fixed you a sandwich, Ellie. I could hear your stomach from the kitchen!" Thinking he is joking with me, I smile and reply, "Thank you so much, Cook. This is so thoughtful. I'm sorry I didn't make it in to greet you yet." Cook blushes, "It's no trouble, Ellie. I know you've had a lot going on." He turns to leave, and I stop him. "Cook, what's the special for lunch today?" He laughs. "Chef has his hands full with the warehouse crew coming back so we are making hot sandwiches to order out on the patio. You can choose steak, chicken or ham, whichever cheese sauce you prefer and then there's a bevy of roasted veggies to add to them. We can make little round sandwiches or large submarine-style sandwiches. I made all the rolls this morning. Nova even helped me chop the vegetables. It should be a beautiful day, too!" And he walks away with a big grin. *If they can move fast, why are all these folks walking?*

I chomp down the sandwich in 5 bites and get back to working on Volley. Now that the warehouse is getting back together, I can look into vacation time for everyone and see who needs to take some time off.

Suddenly, a big bear of a man, almost a lumberjack, comes barreling through the front door. That's the only way to describe it. I don't think there is a gentle movement in this man. He looks directly at me. "Are you Ellie?" I gulp, "Yes, sir. Can I help you?" He grins through the full salt and pepper mustache and massive beard. There's jet-black hair sticking out from under his ball cap. He is wearing jeans, black biker boots, and a red and black checked flannel shirt. There's a little bit of a southern twang to his voice. "Ya, Nathan told me to come to you and get on the schedule."

I am confused but ask what department he is assigned to and his preferred schedule. "I dunno lady, I just wanna truck. That's what I do, and Nathan said I could do it." *Here we go again.... Cleaning up messes.* "It's great to meet you; welcome to the team. I don't have a trucking department set up yet. Would you give me your phone number? I can call you with any questions and a schedule." "Ma'am… I'm gonna stay right here until you have it done. I don't have anywhere else I have to be. And my name is Bryan, ma'am." He plops all his mass into one of the Mission Oak reception chairs lining the front wall. The chair creaks ominously. Nathan comes around the corner, followed closely by Malcolm; it's almost like he heard his name called. Malcolm overtakes Nathan and nearly runs to the burly man. Malcolm embraces Bryan like he hasn't seen him in years. Nathan is standing with his hands on his hips and grinning widely. *Is Bryan crying?*

Malcolm steps away from Bryan, and Nathan steps forward to shake his hand. Bryan wipes his eyes with his free sleeve. Malcolm says, "Thank you, man. Thank you for coming all this way. We really need your help." Nathan opens his mouth, but Bryan interjects, "No man, it was divine intervention. They repo'd my truck last week. I only had the money to get on the bus to get here." And Bryan begins to sob.

Nathan looks around and sees me staring. I divert my eyes to my screen. Nathan puts his arm around Bryan's shoulder and says, "let's go check what Cook has ready in the kitchen. We can catch up over some coffee and a bite." Bryan nods, wiping his tears with his sleeve. The 3 of them walk towards the kitchen. Maia looks over at me and shrugs. I shrug back and turn, once again, to the computer. *It's a wonder I've ever gotten anything done on that machine...*

I hear something… It's getting progressively louder. I turn and look out the front window. A big rig semi truck, shining red and chrome, comes roaring down the driveway. In an instant, it crashes into one of the front porch posts. The entire building shudders. The truck's grill stops directly in front of the window, inches from the glass. I realize that my face is frozen in a scream, and I'm standing up at my desk. I turn to see Maia at the back of the office. *She can move fast, remember?* Nathan, Malcolm, and Bryan are running from the kitchen, followed by Cook, Chef, and Nova. Terrell is coming from the restroom and Mr. Louis is toddling as quickly as he can around the corner from his office.

I see a kid, old enough to drive, but still with a face full of pimples, step out of the truck, look up at it, shake his head, and kick the front tire like the truck did something wrong. In his hand is some paperwork. He comes in the front door, and it slams open all the way around to the front wall of the office. *Gotta get a door closer on that thing.* He walks up to Nathan, "Hey man, I got a delivery for you." *Is this kid drunk?* I've never seen Nathan's face purple. He appears to be barely controlling himself. He grabs the kid by the back of the collar and hauls him out the front door, closing the door behind him at the same time. *My mornings have got to stop being this crazy... Well, YOU wanted a change...*

I hear sobbing. Bryan, the big hunk of a man, is sobbing on Malcolm's shoulder. "How did you do this man? How did you get it back here? Oh, my truck!" Malcolm pats his back. "Don't worry about that now. Let's go finish breakfast." The two of them head back to the kitchen with Cook, Chef, and Nova in tow. Terrell and Caleb are already back in their offices. Maia is standing up at her desk. Mr. Louis is still standing behind Maia with his mouth hanging open. *I didn't think much shocked him at this point.*

I turn back around in time to overhear Nathan yelling at the delivery kid. "Don't you dare step foot on this property again." The kid has enough stones to yell back that he needs a ride to the train station uptown. Nathan blurts out, at the top of his lungs, "Start walking. You can fucking figure it out from there!" He turns, stomping, and slams the front door behind him as he enters. The glass rattles. I don't have the guts to ask any questions right now, and Maia's mouth is also hanging open. Mr. Louis startles and follows Nathan, "Now, Son. SON!"

I see a flash of red from the side window and see my car whizzing around the corner of the building. *What in the world?* It's back to shiny again. That same kid hops out. Today, he's wearing a pair of denim overalls. There are some stains, but they are remarkably clean for having just detailed my car... again. He whistles loud and long, looking at the ruined post and the big truck up and down. He steps in the front door and tosses the keys to me. "Hey, you left the keys in your car again; you sure dirty that car up driving back and forth." I grin, "Thank you for cleaning it for me. Can I get your name?" He smiles a gap tooth grin; there are a couple of pimples showing up on his chin. "I'm Monkey." I laugh. "You mean Grease Monkey?" He smiles again. "Ya.... My real name is Timothy, but it's so *normal*, ya know?" I'm laughing more now. "I hear ya, Tim... I mean Monkey. I'm Ellie." I stick out my hand, but he holds up his glove-covered ones. "I will just get a mess all over you. Do you know if I'm supposed to take that thing?" He points over his shoulder at the truck stuck under the porch. "Monkey, I don't know, and I don't have any authority."

I glance over at Maia. She chimes in, "Monkey, go ahead and take it. I don't know if the truck got hurt, but it sure did a number on the porch." He smiles really wide and jumps up in the air, "YES!" and runs out the door.

I turn to Maia. "I'm not used to crazy mornings like this. Any chance you need a coffee?" She sighs, "That sounds wonderful. You know how I like my cuppa, would you mind bringing me back some? I need to stay next to the phone with all *that* happening." She motions to the mess on the porch. I nod and turn to walk back to the kitchen. I hear a resounding boom behind me as the roar of the big rig's engine winds up. I turn to see what made the noise. Monkey has backed the truck out from under the porch. The entire section of the porch roof has fallen down onto the concrete with a loud boom. *At least it isn't in front of the door.*

###

I step up to the dining room doorway. The 3 guys must have moved on to somebody's office. Cook is the only one still there, but he is holding a large tote of food and starting to step out the door. "Ellie, do you need something?" I nod. "Yes, Cook, but I can get it. We've had a lot of excitement already this morning, and Maia and I feel the need for something stronger." He raises his eyebrows and looks down at his watch. I chuckle. "I mean caffeine, you silly goose!" He smiles. "Ok, I'm heading over to help Chef. If anybody needs anything, send them over there. Enjoy your strong drink!" He walks away, and I walk up to the coffee bar. The options that are available still amaze me. I fix up a coffee with hazelnut creamer for Maia, but I add in a squirt of chocolate syrup to see what she thinks. Then I fix a chai latte again for myself but add a squirt of chocolate to it as well. Crap, I didn't even think to get her a to-go cup. Hopefully, she will be okay with a real mug.

I get back to the front desk and set the cup in front of Maia. She lifts one eyebrow, "What did you do to it?" I grin, "Take a sip and tell me what you think." She sips it, and her eyes open wide. "That's delicious!" "I was a barista at a coffee shop during high school." I blow hot air on my knuckles and pretend to polish them on my shirt, still grinning. Maia says, "You're in charge of coffee from now on." "Yes, ma'am," I reply.

I say to no one in particular, "OKAY, I'm gonna sit down at my desk and get to work on this scheduling!" *Famous last words.* I still need to add Bryan to the system.

Mr. Louis comes around the back of the desks just as my computer fires up. "Ellie! Hey Ellie!" *Oh no....* "Hi, Mr. Louis! What can I do for you this morning?" *Did that sound like I was annoyed? I hope that didn't sound annoyed.* His face is red from exertion. "Mr. Louis, are you ok?" He is nearly panting. "Oh Ellie, you saw this whole mess. I'm just beside myself. I have Bryan's info for you." He hands me a piece of paper with the neatest square handwriting I've ever seen. "Oh, thank you, Mr. Louis. I didn't want to track down Nathan, seeing how angry he is already today."

Mr. Louis looks into my eyes, holding my gaze for longer than is necessary. "You know how stressed he has been. I'm afraid I've left too much on him. I'm so thankful Malcolm is back home. I wanted to check with you. Are you doing ok? The past few days have been a lot." "I'm okay, Mr. Louis, I promise," I reply, but keep my tone even. "I really would like to work on what I was hired to do, though. Do you know if any other employees need to be added at this time?" Mr. Louis looks up and to the left like he is trying to remember something. "No, Ellie, I think Bryan is the only one right now; it was kind of a rush deal." I chuckle, "It sure seemed that way. I will get this entered right away." Mr. Louis is still standing in front of my desk, breathing a little hard.

I stand up. "Mr. Louis, are you ok?" And he immediately slumps to the floor. Maia screams. Terrell and Caleb come running. In mere seconds, they stretch Mr. Louis out on the carpet in the reception area. Maia finds a throw pillow to place under his head. I'm not versed in any CPR or medical procedures. *I can give my cat a pill, but that's about it.*

Terrell looks up at me, "Ellie, could you get a cold compress for his head? He's just passed out. Everything else seems ok." Malcolm, Nathan, and Bryan come running up from the back. They crouch around Mr. Louis, and Maia goes to the supply closet. I run off for a cold compress and some ice from the kitchen. I grab a few bottles of water, too. When I return, Maia has a bottle of oxygen and a cannula in her hand. *Wow... talk about prepared.* Caleb shoos her away. "He's fine, Maia, thank you."

Mr. Louis sits up. "Boys, what happened?"

Nathan has tears leaking around his eyes. "I'm so sorry, Dad, I've stressed you too much today." The color is coming back to Mr. Louis' face. He chuckles, "Did I bang my head? Nathan just apologized to me. Am I dreaming?" That gets everyone laughing, and Nathan's face shows chagrin.

Terrell interjects. "Mr. Louis, could we scoot you over to the wall so you can lean against it?" They move him over, and I hand Mr. Louis a bottle of cold water. He refuses the compress, "It will mess up my makeup." I smile back at him and go back to my desk. *All right. That is enough for today. I'll be the next one to pass out if this keeps up.*

After a good half hour and everyone chatting around Mr. Louis, he asks to try and stand up. The guys all stand in a circle around him to make sure he is steady, and they walk him back to his office.

I shake my head and turn to Maia. She is as pale as I have ever seen her. I rush over to her desk, "Don't you dare. We don't need 2 down today." She looks up at me. "That frightened me, Ellie. I've never seen anything like that happen around here before." I tilt my head, "You've never seen anyone pass out?" Maia shakes her head, "I forgot, you wouldn't understand." She swallows hard, like there's a lump in her throat, and very pointedly returns to staring at her computer screen. *I'm not sure what that's about.*

For about an hour, everything in the office is quiet and some work actually happens. Now it's lunchtime, and I've been looking forward to those hot sandwiches. When I turn around to ask Maia if she is ready for lunch, I see she's gone. And Terrell and Caleb are both gone. *Huh.*

###

I realize I'm still frazzled from this wild morning. I stop at the restroom to make sure I'm reasonably put together. I walk through the warehouse and out onto the patio. There's a long line of folks waiting for food, most of whom I've never met. I see several people staring at me like I'm a pariah. I smile and nod like I know that already. I finally get up to the buffet line. Chef, Cook, and Nova are all standing behind it, fixing sandwiches to order and placing them on plates. These are big sandwiches. *No room for a salad today.* I get to the end of the row, and there are several trays of cookies. Some of them are intricately iced sugar cookies with the company logo; I choose one of those.

I walk past the table with Terrell, Caleb, and Maia sitting at it; they are huddled together, whispering and gesturing wildly. I decide not to interject, so I find another spot at a long table and sit at the corner. It's quite close to the stream, and the sound of the water is soothing but also drowns out a lot of the noise. I eat my lunch in peace, toss the paper plate, and head back to the office alone. The other office folks are still huddled and talking. *Okay... whatever....*

When I'm halfway across the warehouse, I'm stopped by Bryan and Malcolm. Bryan says, "Ma'am, how was lunch? Is it as good as it smells?" I smile and reply, "Sure is Bryan; please call me Ellie." He sighs, "I will try, but I'm not very good at it, ma'am. My southern momma beat manners into us boys." I laugh and start to walk. Malcolm says nothing and only glances at me. They walk away.

###

I'm finally back to my desk, and the voicemail light is blinking on my phone. *Oh no, I didn't even think about setting up my voicemail.* I'm not really sure how to run this phone system so I find a model number on the back and google it for instructions on how to retrieve voicemails. It's a message from Mr. Louis, thanking me for the water and care. *He is so nice.* I sit down at my desk, and I'm GOING to work on the scheduling. *Now, everybody, leave me alone.*

And they did. I didn't see anyone else the rest of the day. I don't even realize it's way past 5 when I stand up and look around to see the sales offices dark. I'm the only person here. I walk back to the dining room to say goodnight to Cook, and it's even darker. Not even a whiff of food. *I will just eat something at home.* But that doesn't appeal to me anymore. My own cooking doesn't compare to what these guys can do. *Maybe it's the mineral...*

I stop at the control panel and shut the rest of the lights off. The only thing still glowing is the security light over the front door. At some point, somebody has cleaned up the mess of the collapsed porch. The night sky is adorned with bright pinpricks of stars. The band of the milky way shines bright. As usual, there is no movement or noise besides the stream babbling. I get in my car and head home. Mr. Mittens meets me at the door, and I grab some cheese from the fridge. I look at my paltry dinner, peel a corner of it off for Mr. Mittens, and head for the tub. Tonight feels especially lonely, and I haven't felt lonely in years. I lay down in bed and collapse until the alarm goes off.

CHAPTER 7:

But it isn't the right time. The sun is blaring through the blinds. The sun should just be coming up. I look down at my phone and realize the alarm wasn't set right. *How did that happen?* I sit bolt upright. *FUCK!!!* I look outside and see the neighbor's cars in their driveways. *Oh... it must be Saturday.* I've completely lost track of this week. *And what a crazy week it has been....* I sigh with relief. Then I remember going to sleep, thinking how lonely I was. *And now I have to eat my own food for 2 whole days. Great.* I get up and throw my hair in a ponytail, slap on some yoga pants and an old t-shirt. I stop at the pantry, noting what I have on hand. I check the fridge, too. Then I look down the basement steps at the pile of laundry. *Some of that is going to have to go to the dry cleaners.*

I start to go downstairs to sort the clothes, and I trip over my own feet. I land at the bottom of the stairs onto the cold concrete floor with a thud. *FUCK that hurt!* I remain laying still for a few minutes, just assessing that all parts are intact, I'm not bleeding, and I wonder if it would be safe to sit up without hurting myself further. Mr. Mittens comes barreling down the stairs and meows loudly right in my face. "Yes, I think I'm ok." He purrs and rubs against my face. It's then that I realize I'm going to have a black eye. I look around to find the culprit and see the smashed laundry basket next to me. The way I landed apparently ruined the laundry basket. It must have broken the fall a little bit, but a corner of the basket matches up perfectly with my tender left eye. *Damn, that's sore. What am I gonna tell everyone at work?*

I slowly pick myself up, only to realize that I've also re-injured my right ankle. The one I broke in college. *Oh, I fell down some stairs that time, too, go figure. I'm gonna be a hot, hot mess come Monday.* I gingerly pull myself up the stairs, one tread at a time. I stop at the top of the stairs to catch my breath, feeling pain all over my body. My head feels heavy. I pause to grab a bottle of water from the kitchen. With great care, I use the backs of chairs and the wall for support as I make my way to the couch. I will need painkillers, an ice pack, food, and probably a pee break... *Let's take one step at a time.* I snuggle under the soft blanket on the couch and fall asleep, ignoring the pain.

###

I startle awake. It's pitch black in the house; I can only see the glow from the stove clock from where I am on the couch. My mouth is parched, and my bladder is screaming. I start to stand and remember a second too late about the weak ankle. I crumble to the floor in a heap. I start to cry. *This is a fine mess.* I decide to crawl on hands and knees to the bathroom, carefully keeping the toes of my right foot from dragging on the floor. Gingerly, I lift myself up onto the toilet and do my business. Wiping is a problem but I get it done. Everything hurts, even my ears. *I can't even get my housework done*. While I'm washing my hands, I find the bottle of ibuprofen and take 2 using the water from the tap. I pull myself along the wall into the kitchen and rummage in the fridge for enough meat and cheese to pull together a sandwich.

I very slowly make my way back to the couch. Getting in and out of that tall bed right now just isn't going to work. Mr. Mittens is on the end of the couch and raises his head to look at me. "Yeah, buddy, you're a big help." I plop back into my spot *(groan)* and turn on the TV for noise. The sandwich doesn't even interest me at this point. I hurt so badly, I'm numb. I nibble here and there to make sure there is something in my stomach. *I'm not sure how many hours I slept, but it sure didn't make me feel any better.* After I finish the sandwich, I throw the paper plate to the floor and attempt to get comfortable around all the bruises and pain.

I wake up Sunday morning feeling marginally better. *I'm going to have to go out for food.* I very slowly toddle back to the bedroom and find clean, soft clothes and the largest pair of sunglasses I own. I carefully change clothes, make my way to the kitchen for my keys and purse, and head out. What I really want to eat is a sandwich from my favorite chicken place, but they are closed on Sundays. I run over to another fast-food joint, and I make it just in time to still get a breakfast sandwich and a cup of coffee. None of it tastes as good as what I've gotten used to at work. I decide to stop at the grocery store, just for items for lunch and supper. I stop at the front desk, and they have last week's dry cleaning ready for me; I now have clothes for the week. *I'm not going near those basement steps for a few days.*

I feel a tap on my shoulder. I turn around slowly and see it's one of my old co-workers. *Oh, this is just great.* I smile as best as I can; it's so painful. "Mariah! How are you doing?" *See? I'm bright and cheerful.* She looks horrified. I must look worse than I thought. "Ellie, are you okay? We didn't know what happened to you. The boss just said you went to another position. Now I find you looking like this. Are you in an abusive relationship? Do you need help?" *She was always so nosey; she doesn't actually want to help, only gossip about me tomorrow at the office.* I sigh. "Thank you for asking Mariah; I am fine, truly. The new job is working out great. I fell down my basement steps yesterday." *Was it just yesterday?* Mariah's smile fades, "Oh, I'm sorry that happened to you. Heal up fast." She strides away. *Yep, no real juicy news here; go on along, little snitch.*

I thank the lady behind the counter and fold the clothes into my cart. I don't want to have to walk back across the store for them when I'm leaving. *They just might have to wrinkle.* I limp through the deli and grab a couple of salads and some fancy lunchmeat and cheeses. I stop in the bread aisle for my favorite sandwich loaf and wobble all the way across the store to the dairy aisle for some creamer for my coffee. By the time I make it up to the register, I'm exhausted. The cashier, a teenage boy, blurts, "What happened to your face?" I mumble that I got into a fight with a mountain lion, pay for my food and get out of the store as quickly as I can.

###

I'm limping across the parking lot when I see a very large, lifted, and loud white Ford diesel pickup truck heading straight for me. I cannot move out of the way fast enough. But it stops on a dime right in front of me. I look up, and it's Chef. He jumps out of the truck and yells, "ELLIE, what the FUCK happened to you?!" I'm mortified. *I really just want to go home and hurt by myself before I have to face the people at work tomorrow.* "Hi Chef, I'm okay. Really." He cuts me off, "You are not okay; you can't even walk." At that, he scoops me up into his arms, my purse dangling from my shoulder, and very gently places me into the front passenger seat of his truck. He grabs my bags and clothes out of the cart, sets them in the backseat, and starts driving.

I try to protest, "Chef, my car." He interrupts, "We will come back to get it later. They have 24-hour security detail here, it will be fine." I am so exhausted and embarrassed I decide not to fight him on it. "Mr. Mittens…" I say before I pass out.

***

CHAPTER 8:

I found all of this out later: I'm still passed out. Chef pulls right up to the front of the office, lifts me out, and carries me through the front door. Nathan stops him in the hallway, "What the FUCK Chef?" "Look at her face, Nathan. I don't know what happened, but it has to be painful." Chef replies in a huff. Nathan leads him to the infirmary. *I didn't even know we had an infirmary.*

###

I wake up; I dunno when. I'm in a hospital bed in a room filled with all manner of healthcare equipment. There is a nurse in bright purple scrubs who walks up to me when I try to lift my head. "Now, dear, you keep your head down on the pillow. YOU have a concussion." *That explains all the sleeping yesterday.* I'm not in pain, but I sure feel lousy. *They must have given me some heavy pain meds; those usually don't agree with me.* The nurse picks up the handset on the wall phone and tells someone I'm awake.

Nathan strides in, Chef following close behind. Chef smiles really big and says, "If you wanted attention, all you had to do was say so!" He very tenderly pats my shoulder and mumbles something about getting better and that he has food to cook. As Chef walks away, Nathan walks up on my other side. I try to turn my head, but everything feels woozy. "Ellie, why didn't you call when you got hurt?" Nathan starts in on me.

Mr. Louis barges into the room, "Nathan, you leave that girl alone. She has been through enough the last few days." *Last FEW days?* I try to talk, but it comes out as a whisper, "What day is it, Mr. Louis?" "Girl, it is Tuesday." He replies. *TUESDAY?!* I start to fling my blankets off and sit up.

Nathan gently pushes my shoulders back down. "Stay put; everything is fine. I'm just thankful you weren't hurt worse." I'm floating off to la la land again, "Mr. Mittens…"

###

I wake up and feel normal, like normal normal. Nothing is particularly painful, and I can move everything without it screaming in pain back at me. I'm still in the hospital bed. Mr. Mittens is lying on the foot of the bed, purring. I start to sit up, and the same nurse, this time in blue scrubs, says, "Take it easy, Ellie. Just sit up slowly. You're healing, but things are still weak."

I take her advice and slowly sit myself up in the bed. She raises the back of the bed to match my angle. "You have to be hungry. Chef made you some chicken noodle soup from scratch. It's the best." She says and sets a very full, very large bowl in front of me on the tray table, along with a sleeve of saltine crackers. My arms feel heavy, and my face feels tender. I turn my head from side to side, and the creaking in my neck is gone. I smile back at her, and it doesn't hurt to do it. "I guess I hurt myself worse than I thought." The nurse replies, "Oh yes, Nathan said from the shape the washbasket was in, you had to have cracked your head on the wall." *Nathan was in my HOUSE?! Oh yeah, the cat.*

I set to eating the soup; I cannot resist the delicious aroma any longer. Cook walks in, "There she is! Miss America!" he sings. I smile back at him. "Your brother beat you to feeding me today." Cook laughs, "His chicken soup is better than mine; he makes the noodles homemade. I refuse to make homemade noodles." I laugh and take another sip of the soup. I don't even need the crackers.

I have a thought, "Cook, what day is it?" He screws up his face, "Now, Ellie, don't be mad. It's Tuesday." *TUESDAY?!* "You've got to be fucking kidding me. Fucking TUESDAY?!" I yell back at him. He doesn't deserve to be yelled at, but how did I lose a week? No, I missed out on 9 days. *NINE DAYS.*

Cook smiles but looks afraid. "You needed to rest and heal Ellie. You took quite a spill. I'm not sure what would have happened to you if Chef hadn't found you at the grocery store. You shouldn't have been driving, let alone out and about." I shake my head. "I need to get home and get the housework done and the laundry. I don't have any clothes, and I want to get back to work."

Cook grasps my hand. "Everything is already done, Ellie. Your house is clean, your fridge is stocked, and your laundry is done. Nova and I did it all yesterday." *How am I going to repay everyone for this? What about the medical bill?*

Nathan and Mr. Louis walk in. "Oh, thank goodness, you're looking so much better, Ellie!" Mr. Louis kisses my cheek. *What the?* Nathan walks up but has one hand behind his back. "You have to guess what I brought you." *Not flowers, not flowers, not flowers.* "No, not flowers; I know that's what you're thinking. Here is a chai latte and a tablet that has Volley downloaded on it." I smile, touched. "Nathan, you knew exactly what would make my day."

The nurse instructs me that she wants me to remain in the infirmary for another day before trying to sit up at my desk to work. She also relates, "You're only going to be able to work half days or less for at least a week before going back to 8 hours in front of the computer." *Lady, I haven't had 8 hours in front of the computer yet.*

Maia walks in, gently knocking on the door frame before barging in like the guys. "Ellie, I'm so glad you're improving. We have all been so worried about you. I miss talking to you at our desks all the time." She hands me a big bar of chocolate. Nathan, Mr. Louis, and Cook wave their goodbyes. I turn back to the nurse, "How badly was I hurt? It had to have been more severe than I realized for it to take 9 days away from me." The nurse replies, "the MRI still shows some bruising on your brain. You've been at risk for a stroke since the initial concussion. Falling in and out of consciousness is an indication of a nasty concussion." *MRI? This bill is getting bigger.* I ask, "Where and how was I transported for MRI?" She replies, "We have a full-capacity, state-of-the-art medical facility here. There are 2 more rooms, just like this one, across the hallway and a surgery suite around the corner. If you need something done, we handle it all in-house. I'm thinking you're worrying about a medical bill?" I nod and continue to eat my soup. Maia is standing next to me, patting my hair. *Everybody sure has gotten affectionate this week.* The nurse continues, "We handle everything you need medically. You don't have to leave to see any physicians. If there is a specialist needed, like in this case, the guys have contacts all over the country. You were evaluated by 3 top neurosurgeons, 2 in person and one using your chart. There is no bill; it's a benefit of working for the company." *Holy shit.*

I'm flummoxed by all this information and stop eating. Maia is smiling down at me. "Well then, I had better get to work." I fire up the tablet Nathan brought me. Mr. Mittens is content at my feet. He seems content with everyone that's been in the room; the stranger-danger he had before has resolved.

Maia steps away, "I'm heading back to my desk. When you're ready, holler at me, and I will get you settled in your apartment." *WHAT IN THE ACTUAL FUCK?!* And she walks out. Leaving me here with these thoughts, all by myself.

###

I stay the rest of the day in my hospital bed, save for potty breaks, and work on Volley. Once the workday is over, I have a barrage of visitors. Every one of them just stops to say hi, so they say. *Um, Ok...* I smile and make small talk with each person in turn. Finally, about 9pm, Nathan walks in. With Malcolm in tow. *Great.* They both pull up chairs and sit at the foot of my bed. Mr. Mittens leaves me and goes to settle in Malcolm's lap! *Traitor.* Both Nathan and Malcolm smile and wink at each other. I look back and forth between them and growl, "What?!" Malcolm opens his mouth to speak, but Nathan interrupts him with a look. Now I'm getting pissed. "What?!" Nathan starts, "It's time, Ellie. You need to know it all. And you will need to make some decisions afterward. The doctors have all been amazed at your recovery." Malcolm and Nathan exchange a look again. Nathan continues, "I need you to stay calm, set your emotions aside and listen to the facts. Then you can make an educated decision." I agree, even though I'm not sure I will be able to control my angst.

Nathan starts again, "You remember the other night when we started our conversation? About the mineral and it's benefits, how the business started?" *That night seems like a lifetime ago...* I nod and he continues, "The mineral is far more powerful than we are allowed to tell the common public. That is why the usage rates set by the government are so low. I told you that we use a higher dosage here, in our food and drink, even in our water." I nod and he reminds me, "You seemed a little aggravated about not having a choice to take the mineral." I nod again, a bit more vehemently. "The doctors decided the best option to heal your brain was with a drug-induced coma. You had severe bruising on one side and it was going to take time to heal the damaged tissue inside your head. The other option was surgery and I knew that would leave you with a deficit for the remainder of your life." *Who knew falling down the stairs could cause such an injury? Oh no, what if Chef wouldn't have found me? Would I be dead?*

I begin, "Nathan, I..." He interrupts me, "Please Ellie, just listen. Yes, you would have been dead by the end of day Monday if Chef hadn't found you. The doctors know about our nutritional supplement, but they aren't aware of it's power. I know you don't like having your choices taken away. I made a pile of decisions for you that I would have preferred you would have been able to make yourself. I lied and told the doctors that I was your brother, so I signed the documents to put you into the coma so you could heal. You had to have an IV in your arm to give you medications, fluids, and they put in a temporary feeding tube while you were asleep." He interrupts himself, "Bear with me; I will come back to that." I nod, and he continues again. "Usually, when folks join our team, they are fully educated on the mineral and what it can do. They understand the changes it will make to their body. They also understand the risks of someone outside our organization finding out and what a shit-storm it would cause in the media. Also, what the general public would do to our lives and livelihood here." I tilt my head in puzzlement. Nathan holds up a hand, "Let me finish. "Since you already had the IV in your arm, I talked with our laboratory lead, and we created a concentrated solution that would go through your IV using the regular saline. While the mineral is nearly miraculous in what it can do to heal and restore, it isn't instant. That is why you still have a little bit of bruising in your brain and you're taking it easy. If you notice, your bad ankle is healed completely." I turn my ankle and notice that even the achiness it has had since college is gone. My mouth forms a little o.

Malcolm interjects then and his deep voice absolutely makes my ears sing, "Ellie, when Nathan says that people know what the mineral can do, they understand that it will take months to heal everything they have going on in their bodies. They know that their life is completely upended and they are living in our on-grounds housing. Nathan did not follow our usual protocol with you. That is why Dad and Cook have been up his ass about it." He smiles with a little bit of satisfaction and Nathan punches him in the shoulder. Malcolm continues, "Nathan is trying to explain that to save you from a life of disability, he chose to do an un-regulated and un-approved treatment to you, without your knowledge or consent. Also, upon your first interview, he should have explained the secrecy and ramifications of belonging to our family here. He also should have forced you to move onto the grounds. His workload has been far too great and Dad and I have failed him. We failed to see he was in trouble and failed to step in sooner to assist with all the demands on him." Nathan looks over and Malcolm and smiles a little smile.

I sit up a bit straighter and open my mouth to speak. Malcolm cuts me off. "Ellie, the last 9 days have been excruciating for all of us. We didn't know if our treatment would work. The doctors are calling you a miracle. They don't know what Nathan did and we know it will stay that way. We also hope you don't hold the choices that he made against any of us." I nod and gulp a big breath. "Everyone here at the business knows your case and knows what happened. They all already knew that protocol wasn't followed at the time of your hire. Would you be willing to help us remedy that one fact? Maia has already prepared one of the apartments closest to the office for you. That way, you can commute to work with a short walk each day instead of trying to cover any distance." I interrupt, and he holds up a hand, "for just a couple weeks, until your brain has healed completely. You haven't had any time to research your injuries or condition, but you need to know that normally a person with this severe of a head injury, if not dead…" He stops, gulps, and looks at Nathan, "Will have a permanent disability for life, as we discussed. But the damage to your bones and joints would have taken months to years to heal. And like your previously broken ankle would have ached the rest of your life. The human body can heal, but it remembers what hurt it. The mineral cuts through all of that and gives each piece of cartilage, every bruise, even the marrow in your bones, what it needs to fully purge the damages caused by everyday life. It's truly a gift from God." My eyes must be like saucers. Nathan smiles and stands to go over to the cabinets across the room. He finds a hand mirror and brings it back to me. He holds it up to my face and says, "Look."

My skin! My skin is like alabaster. It's completely smooth; my eyes are a glowing almost turquoise. I decide the most telling factor is going to be my teeth. They are bright white and straight. *Oh my.* I open my mouth to look at my bottom teeth. I glance up at Nathan, "Your snaggle tooth will repair over time." And he chuckles. "You're going to notice that every scar and bump is gone. Your hair and nails have already grown quite a bit. I'm going to venture a guess that after you heal the bump inside your head that you're going to feel you think more clearly, your memory is functioning better than ever and…" Malcolm interrupts him, "usually this process happens over months for a new hire. Sometimes, things don't change for a year or two. It just depends on the damage inside your body that needs repaired." I'm tired again. I look up at Nathan who is still grinning at me, standing over the bed. "Malcolm, I think we've shocked her enough for one night; let's let the woman sleep."

They both bid me goodnight. Malcolm lifts Mr. Mittens back onto my feet, and I fall into a dreamless sleep.

I know I've lost track of days vs. nights and what day it even is. I wake up and cannot tell if it's dark outside or what. The nurse comes in, "Morning dear." I startle as I was apparently lost out in Lala Land. I look at her; she is in bright pink scrubs today. I love that she is always so put together. Her hair is long, brown and she has green eyes. I cannot tell if she is particularly tall, but her cheekbones are high and she has a prominent nose. "Hi! Uh, I'm sorry. You've been taking care of me all this time and I don't even know your name." She smiles. "Honey, I'm Kathleen. I married to one of the salesmen. I'm on call 24/7 to come help, but usually, I'm a stay-at-home mom to 4 under 4." She smiles wide at the shocked look on my face. "You're just fine dear. You've been very ill and needed to heal. That was quite a tumble you took!" I'm chagrined, "Thank you Kathleen, for being here. Would you tell me about your kids? Who has been taking care of them since you've been taking care of me?" Kathleen smiles and fills me in on each of the kids and their ages, likes and dislikes. "You've met Amelia?" I nod, "She seems nice; the one time I met her." Kathleen interjects, "She is great with the kids and loves taking care of them. You know she is almost ready to pop." I'm shocked; I knew she was pregnant but didn't realize she was that close to her due date. *Amelia has to be a superwoman too, taking care of herself, Caleb and now Kathleen's brood.* "Would you like some breakfast? You have to be hungry by now." I nod and ask, "Kathleen, how am I going to repay everyone for all this?" Tears start leaking from the corners of my eyes. She walks over and pats my arm, "You don't dear. When somebody needs your help, you are just ready to step in to do whatever is needed. Just like you did with Mr. Louis the other day. Nathan was right about you being a perfect fit here." And she glides out of the room. Mr. Mittens stands up, stretches out full with his butt in the air, tail high. Then jumps down and trots off to follow her. *I wonder if they have gotten used to an animal around here. Oh no, who has been cleaning his litterbox?*

Cook brings in my breakfast. "Cook! You didn't have to deliver this. I'm supposed to be allowed to get up and do things now." He grins, "I had to come check on you. You're looking fresh as a daisy!" And his face turns red. "I knew you would want your usual." And he winks. He lifts off the plate cover and reveals a mounded plate with all the different meats, potatoes, toast and this time there's scrambled eggs. "I didn't think the over-easy eggs would hold up to the travel." He stays, makes small talk, and watches me polish off the entire plate full. Then takes the dishes back with him.

A few minutes later, Maia does her little knock on the door frame. "Girlie! You ready to move on out?" I stammer, "I, I guess so…" And she helps me stand, makes sure I'm steady and grabs a duffle bag from under the desk across the room. She stuffs all the clothes into the bag, walks out, finds a cardboard box, and comes back to place all Mr. Mitten's things in the box. "I will send Terrell out to grab Mr. Mittens things. I'm going to carry the duffle bag." She hands me a cane. I look at her, indignant. Maia sighs, "Just use it. I know your body is healed, but your brain is not. I told you the other day, I can't handle another one passing out. I will feel a lot better if you have something besides me to hold on to. She has the duffle bag slung over her left shoulder and curls me into her right side. She wraps my left arm around her waist and says, "Now hold that cane in your right hand. We are going to walk slowly over to the ATV."

CHAPTER 9:

We step out of the infirmary and I realize that it is right across from the bank of protective clothing and supplies for the harvesting crew. She settles me into a shiny red ATV on the passenger side. Maia puts the duffle bag into the bed of it and walks over to grab the keys from a hook on the wall. "Off we go! Hang on!" She squeals and she takes off. I wasn't prepared for how fast she was going to drive, but I notice she turns right instead of left like Nathan did on the tour. We drive as fast as the ATV will go down a dark hallway and stop abruptly at a set of double all-glass doors. The whole trip only takes a few minutes. She hangs the key on the wall next to the door and stops to help me out of the ATV.

Inside the glass doors is a carpeted, well-lit hallway lined with plain black doors on either side. It's very quiet. She turns to the left and stands me in front of a door marked with a large gold B. There's a peephole in the front and no lock on the door, just a handle. She turns and says, "Welcome home. Well, atleast for a little while." Directly inside the door is what I could only describe as a high-class hotel suite. To my right is a fully functional but efficient kitchen with cherry wood cabinets and black granite counters. The appliances are all stainless steel. On the left is a closet and hall tree that would be a great place to drop your purse at the end of a long day. Then, the whole place opens up to a large living room with 2 couches facing each other, a large fireplace at the end, with a big TV over the firebox. She guides me through the living room, turns me to the right, and there's a closed door. She opens it and steps inside. *That is the biggest, fluffiest bed I've ever seen.* The lamps are on, and the whole room is bathed in a golden glow. All the furniture is a whitewashed gray, the walls are a cream color. The bedding is all ivory cotton. My mouth is hanging open.

"Check this out," she says as she takes my hand and leads me to the right. She guides me into a white marble bathroom. There's a jacuzzi tub in the middle of the room and a double vanity on the left. Beyond that is a set of folding doors that could only be a closet, and to the right is a single door. Maia says, "Yep, that's the potty." And giggles like a schoolgirl. I continue to look around and see a wall of glass behind the tub. She points and says, "go have a look." It's a marble shower, 2 rain shower heads on the ceiling and another set of various nozzles and handles all over the wall. "Maia… I… don't need all this." I look up at her and start to cry. She guides me to the corner of the tub. "Sit. You are going to have to come to terms with everything and that is going to take time. Let's get you settled so you can rest and watch some TV. I'm sure there's some shows you've missed watching." She helps me stand and I wobble back out to the living room. She gets me settled on the far couch with a fluffy blanket and cup of tea.

There's a knock on the door. I hear a cat screeching and growling. *Oh dear…* "Mr. Mittens! Ow! Damn cat!" I hear Terrell's deep baritone voice outside. Maia is grinning from ear to ear. Terrell steps inside and, holding the cat by his scruff, says, "Missing somebody?" Mr. Mittens takes another swipe at Terrell's arm and Terrell lets go. Mr. Mittens streaks across the floor and hides under the couch I'm sitting on. I sigh. *Atleast he is close by.* Terrell sets the cardboard box on the kitchen counter. "Here's the lion's things. Good to see you Ellie." He waves at me and kisses Maia on the cheek before stepping out the door. *Uh-huh...* Maia blushes. "I guess you know about that now too." I laugh, "Maia, I've known about that from the first day I was here." She grins, "Oh all right. Do you need anything before I go? I need to get back to the front desk." I shoo her away, "No Maia, I'm fine. I will be careful and try not to wobble." She glares down her nose at me, "Promise?" "Yes ma'am, I promise." And I smile back at her. She steps out and gently closes the door. "Mr. Mittens, Mr. Mittens, here kitty kitty!" I call. Nothing. *Oh well, he liked people for a few days...*

I sip my tea and start thinking. *Dangerous, I know.* I'm definitely in over my head on how far this mineral thing goes and how it is distributed all over the world. I'm sure it would also cause a complete destruction of the source if the general public understood its power. *People as a whole tend to destroy anything that is good unless it is severely protected.* My whole world has just been turned upside down. Well, mine and Mr. Mittens' world. The folks here are all amazing. *When Kathleen said well chosen, it wasn't just me that was well chosen.* I mean, how do you add so many caring, well-educated, willing people? *The normal interview process must be extremely in-depth.* I find myself getting tired again. *I wonder how long it will take for this fatigue to lessen.* I turn on the TV and find something inane to fill the noise in the background and snuggle down in this super soft blanket. I fall asleep.

###

I wake to a knock at the door. "Ellie? Is it ok if I come in?" It's Kathleen. I yell back, "Of course! Please come in!" She is carrying a box, what looks to be a heavy box but it sure isn't fazing her muscles. I stand up and meet her in the kitchen. "What is all this Kathleen?" I ask, marveling at the food-filled box. "It's all the food from your house. Cook just ran over and boxed it up, he was worried it would spoil before you returned home. He asked me to bring it by because he is in the middle of the lunch rush right now." *Is it noon already?* "Are you hungry dear? I could fix you something quick." She is always so thoughtful. "Oh no, Kathleen, I drank some tea a little bit ago. I'm not ready for a meal yet. Thank you for bringing this." I motion to the box. She stares into my face like she knows it by heart. "Ellie, you're no longer under my watch, but you have to promise me that you will take care of yourself." I cower under her gaze, "Yes ma'am." I feel like a child that has done something wrong. She chirps cheerfully, "That sounds good. Now let me get all your vitals for the doctors. They are just so amazed at your recovery!" She winks. Mr. Mittens comes out, led by his nose. "There you are, Mr. Mittens!" She walks over to pet him. "We became good buddies over the past week or so. So nice to have an animal around here." "Hey, Kathleen. I…" She cuts me off, "Now that you're fixed up, I'm heading home to my kiddos. You continue to heal Ellie. I will be back in a day or so to check on you." She leaves and I start to put away all the groceries, deciding where everything should go.

There's another knock at the door. I holler, "Grand Central Station." There's a deep chuckle on the backside of the door. *That voice makes my spine tingle.* Malcolm. He peeks his head in. "You're up and going. That's a good sign." He leaves the door open, "I just wanted to stop by and bring you these. I know you're a country girl." It's a bouquet of peach roses arranged in a blue glass jar. They take my breath away, "Malcolm…. I…" "Shhhhhh," and he puts a finger to my lips. *That left a zing.* "You should have had them sooner."

Mr. Mittens rubs up against Malcolm's leg. Malcolm bends down to swoop him up. "Crazy cat. You know how much this cat loves you, right? He was heartsick when Nathan went to the house to get your clothes. He grabbed him and brought him along. Mr. Mittens didn't leave your bed except for necessary functions the entire time you were asleep." "Malcolm…. I…." I start to reply, but I hear Malcolm say under his breath, "Awwww fuck it." He leans in and brushes a kiss across my lips. *I was not prepared for that.* I grab for the edge of the counter. "Ellie, I'm sorry. You're not steady on your feet yet. Let's get you settled back down." He leads me over to the couch and covers me up with the blanket. The front door is still open, and Mr. Mittens takes off down the hallway. I point to the door, "Malcolm. The cat." He yips, turns, and runs out to the hallway. Mr. Mittens didn't get far with as fast as Malcolm can move. *At least he likes Malcolm. You could do worse….*

Malcolm steps inside the door, deposits the cat, and waves. He shuts the door behind him this time. I grin and press a finger to my lips. They are still warm from that little kiss. I can just see the tops of the roses from my spot on the couch. I'm still smiling as I lay back down.

CHAPTER 10:

I wake up ravenous. I don't know what time it is. I get up and walk to the kitchen to find my purse. *Wow, I'm better. I didn't even think about getting up and moving; I didn't even sway. What a relief.* I cannot find my purse. *Where is the thing?* I forgot about the duffle bag. I walk to the bedroom. The duffle bag is sitting on the bed. I stop and peek my head into the amazing bathroom just to marvel at it. I dig to the bottom of the bag and find my purse. *Whew.* I search out my phone, and it's dead, not slightly dead, but completely dead. *Where am I going to get a charger for it at whatever time this is?*

All of a sudden, it hits me. What about the bills? *I've been down for 2 weeks now. Or so?* Nothing has been paid for at least that long. I haven't submitted any paperwork to have my paycheck deposited at the bank. I can't even think of what would have been due in that amount of time. I sit down on the edge of the bed and just breathe.

I hear a crash. *Oh no, the roses!* I stick my head out of the bedroom door and see Mr. Mittens' tail go just out of sight under the couch. The door opens, and Nathan steps in, whispering, "Ellie? Ellie? Are you asleep?" I can see the tops of the roses over the bartop to the kitchen. *Whew.* "Hi, Nathan, I'm over here. You can come in." He walks into the living room and is carrying a tray of food. It smells delicious. "Cook sent this with me. I stopped to see him before I came to see you. I think he wanted to bring this to you himself." He points out a tiny bud vase on the tray holding a single daisy. "Oh, Cook is so thoughtful," I reply. *If Nova tames down, she is going to have herself a good one.*

Nathan sets the tray down on the coffee table and looks down at the floor. "What happened here?" There's broken pottery scattered across the top of the coffee table and out onto the floor. *That was the crash.* "Well, Mr. Mittens doesn't like you." I'm cut off by a resounding guffaw. "That's the understatement of the year!" Nathan breathes in between laughs. "You should have seen him on the ride from your house to the infirmary. If looks could have killed, I would have died going down the road. He didn't even appreciate being driven in a Jaguar! Shotgun rider even!" And he continues to laugh. "It's so good to see you laughing, Nathan." I'm smiling up at him. That sobers him. "Yes, I've needed to ask for help around here far longer than I would like to admit. Having Malcolm back home and everyone's support has helped me to sort out a lot of the problems I caused."

He changes the subject, "Hey Ellie, do you like your apartment? Do you need anything?" I shake my head. "Nathan, this is more than I could ever need. I don't know how I'm going to repay everyone for everything." I interrupt myself this time, "Actually, Nathan, I do need something. I need a phone charger for this kind." I hold up the butt end of the phone. "And I need to find the bills at my house so I can get them paid." Nathan steps back 2 steps. "Now, Ellie, don't be mad at me."

I roll my eyes. *I think I hear my eyes roll; they go so far back into my head.* "What now?!" Nathan is backing towards the door. "Ellie, I knew you would be out for at least a few days. When I went to your house to gather things, I saw your planner sitting there and grabbed it. Your bills are paid." His hand is on the door handle. My face must be showing murderous tendencies. "NATHAN MICHAEL LOUIS!" *I dunno if that is even his real middle name, but it sounds good coming out in a mom-yell.* He is out the door, and it's shut tight, all in the matter of a second.

I peek my head out, and he's gone. I see Terrell's head sticking out of his door further down the hall, and all the way at the end I see Malcolm peek his head out. He steps out and walks slowly this way. No shirt on, just gym shorts, the gray ones that hang just right. *I'm gonna hyperventilate.* "Ellie, you ok? I heard you yelling at Nathan. His middle name isn't Michael, ya know." He's grinning big, really big. He is about 30 feet away now. Before I fall down, I smile, say I'm fine, thank him, and duck my head back in, slamming the door. I can't handle tall, sexy, half-naked man right now. *Nathan always stresses me out. And Malcolm turns me all the way on..... Fuck. Yeah, that's what I wanna do to him too. Okay, get off this train.*

I distract myself with the delicious supper Cook sent. Then I decide it's time to figure out the fancy shower. I strip off and start turning knobs and pushing buttons. *Oooh, it's a steam shower!* There are even colored lights that set off at different intervals, like you're having a disco in the shower. *That's a bit much.* I find the button for the lights and turn those off. With all the hot water spewing out of various places, I'm clean in a minute, it seems. I stand in the hot water for a few more minutes just to enjoy it. *How did I get clean while I was asleep all those days? Best not to think about that.* I step out and towel off. It feels really good to be clean.

I dig through the duffle bag, mortified when I find an entire pocket filled with bras and panties. Nathan must have brought those, too. I would have rather gone commando before having my boss root around in my lingerie drawer. I put on an old T-shirt I find in the bottom of the bag and slip under the covers on the bed. It's a pillowtop. I sink into the soft mattress; the blankets are fluffy, too. I drift off.

I wake up to an alarm going off, but it's far away. That's my alarm! I throw off the thick blankets and try to get into the kitchen to shut the thing off. I catch my leg on the corner of the coffee table and go down. *Not this again.*

Just then, Kathleen pushes open the door and calls, "Ellie? Where are you, Ellie? I have to do your vitals!" She steps over to the kitchen to push the button on the phone for the alarm. "Ellie?!" She finds me next to the coffee table, groaning. "Now, Ellie, you promised me you would take care of yourself. Am I going to have to wrap you in bubble wrap?" I roll over and look up at her. "Kathleen, I've always been a klutz; it isn't going to stop just because I've taken a nutritional supplement." She laughs and helps me stand up. "Let me see it." I look down and see a bruise has formed but already turned green, right above my ankle. She tenderly touches it, and I recoil. "Ouch, okay. Let me get an ice pack." I sit down on the couch, and she brings the ice pack; it feels great on the bruise. She starts taking all my vital signs. "I'm going to have to wait a few minutes and take your blood pressure again. This one would get us both in trouble." She shows me how high it is. "I'm sure it's just because you fell down and are running adrenaline. Do you need anything?" I shake my head no.

She visits with me about her kids and all the drawings Amelia had them do while Kathleen was gone and taking care of me. "They are so talented, my crew. They even drew a picture for you!" She reaches down and pulls it from her bag. The corners are wrinkled, but it's a drawing of a field, with the sun up in the corner and puffy clouds. There's a single daisy in the field of green grass. And I start to cry. "Thank you, Kathleen and thank the kids. I'm looking so forward to meeting them." After 15 minutes or so, I've calmed enough from my fall to have normal blood pressure. Kathleen heads out the door, promising to be back tomorrow. *I wanna be at my desk tomorrow.* The bruise is fully green, the ice numbed it well. I decide I'm done with the ice pack and put it back in the freezer.

Now to see about that phone. I find the phone plugged into a charging cable, which is plugged into a wall socket behind the coffee maker. *Nathan must have snuck in and done that while I was asleep. I will have to thank him when I see him next.*

Now that I know it's 10 am, I decide to dress and fix a bite to eat. Eggs sound fine, maybe some toast. I don't know how to get back around to the dining room from here yet, so I will just stay put for the moment. Next to the stove, along with the standard salt and pepper shakers, is a white shaker, unlabeled. *I bet I know what that is…* I sprinkle some on the eggs while they cook. I was right; these are the best eggs I have ever made myself. I find a toaster in a base cabinet and plug it in. Inside the box of food is a loaf of my favorite sandwich bread. Cook thought of everything. I find butter in the fridge. *I've got food in me; now I have to find some decent clothes.*

Avoiding the coffee table, I walk back to the bedroom and open the closet. Nothing. I remember there's a closet in the bathroom. I walk back around to it, and there hangs my dry cleaning from the grocery store debacle. I've got a couple pairs of pants, a skirt, a couple blouses, and a sweater. *I can work with this.* Since I don't go for any bold colors, most everything coordinates, even if it wouldn't be my normal color coordination. *Shoes are going to be the issue.* I'm sure heels aren't going to be approved after this kind of injury. I know I have some flats in my car. *Oh, where is my car? Monkey probably went and picked it up.* I dress in dark green slacks, a light pink blouse, and an oatmeal-colored sweater. *These are old lady colors… oh well, the best I've got for right now.* I remember the black ballet flats I stuck in my purse for emergencies. *Hallelujah! I have shoes.* I don't have any perfume or makeup to put on, and the provided toiletries are just generic, but I'm clean, dressed, and have covered feet. I pet Mr. Mittens, who has snuggled into the thick blankets on the bed, and grab my purse and phone off the kitchen counter. I leave the apartment and I remember being right next to the exit doors. *Well, atleast the set I know about so far.*

###

There is no ATV sitting outside the double glass doors. I start walking down the dark and truly dank hallway back the way I remembered traveling the other day. I dig my phone out of my purse to use as a flashlight. It's very dark. And it's far longer than I remember. I've been walking for about 10 minutes when an ATV with the lights on barrels up to me. It's Terrell. "Ellie! I'm so sorry. I should have left this for you this morning. Normally, I go out of the apartments in the other direction, but today, I needed to visit with Chef." I smile at him and thank him for picking me up. *How did he know I was here?* "Terrell, you mentioned another way out of the apartments?" I say. He replies, "Yes, we should have shown you, but we were so intent on getting you settled and comfortable. If you go down the hallway to the other end, the hallway is nice and lit and curves back around to the office." *Well, gee, that would have been good to know.* Terrell replies, "I know, Ellie; I'm sorry for that, too." *What the?* I shake off that suspicion and visit with Terrell about little things.

Then I bring up Maia. He blushes; you can just see it high on his cheekbones. "Ellie, I need to tell you a secret." "Okay, Terrell, I have no one to tell. What's up?" and I look over at him. "I went out last week and found a ring." He smiles that mega-watt smile. "I'm so happy for you, Terrell!" I reply, smiling back at him. He seems genuinely happy. *I know Maia will be.* Terrell drives right up to the door leading to the main hallway at the office. I step out of the ATV. "I'll see you later on, Ellie." Terrell smiles. I wave to him as I grab the door handle, "Thanks, Terrell!"

I step inside the office and turn to my left. Cook is putting the finishing touches on lunch. Nova is behind the counter, too. *Does this woman sleep?* "Hi Cook! Hi Nova!" Nova runs around the counter and barrels into me, wrapping me up in a tight hug. *Whoa.* "Ellie, I am so glad you're ok. I'm sorry I haven't been very friendly to you." *Okay, what happened here?* She looks me up and down. "I think we should call you Mildred today instead. Does that sound more normal?" She laughs and walks back around the counter. I laugh, too, and ask, "What's for lunch today?" Cook chimes in, "Your steak salad you like. I was hoping we would see you today." *How did he know? It can't be....*

I sit down at the end of the bar, and Nova brings me a plate full of that glorious salad with tender strips of steak splayed over top. She puts a bottle of Cherry Limeade soda in front of me, too. *That's a blast from the past.* She explains, "Kathleen said you were talking in your sleep. Kept saying something about Cherry Limeade." My face blushes deep red. *I don't talk in my sleep. Surely, I don't. I bet I don't snore, either.* Nova gives me the side eye and laughs. "Eat up, you need your strength." She looks up. Within a minute, the entire office is inside the dining room, and everyone has something to say. *Everyone except Malcolm.* I realize. I smile and nod to everyone and say hello across the room to others in between bites. I'm not the only one who ignores table manners with this delicious food. All of us are scarfing down the meal as quickly as we can and asking for seconds. *Like a pack of wolves.* Cook laughs. "There's plenty more; ya'll act like you didn't get breakfast!" Finally, the assemblage of bellies are full and finished with dessert. One by one, we all leave to find our respective places. *That was eye-opening.*

✻✻✻

CHAPTER 11:

I stop in the restroom and find Maia sitting in the corner, on the floor, crying her eyes out. Mascara has dripped all down her face. I open my mouth to speak and she holds up her phone; the headline reads, "Quapaw Nation Reservation in Oklahoma consumed by wildfire. 3 Elders dead." She is looking up at me, her eyes shining. "Grandpa" is all she can say. I'm as devastated as she is, and I've never met the man. I sit down next to her and reach over for some toilet paper for tissues. We both cry. Her for her family and me for my independence.

Terrell comes rushing into the ladies' room, the front door bangs against the wall. "Maia! Ellie!" He picks Maia up and holds her tight in a hug. "I'm so sorry, Maia. What do we need to do? Do we need to go to Oklahoma?" She nods, and he walks her out of the bathroom. I stand. Nova walks in, "Ellie? You ok in here?" "Thank you, Nova, I'm okay." I sniffle. "Did you hear about Maia's family?" I ask. Nova nods solemnly. We walk out of the ladies' room together. She heads back to the dining room. *Maybe I should switch her department from cleaning to kitchen help.*

I walk around the corner into the office. Outside the front door sits a jet-black Camero. Monkey is just getting out of it. He hands Terrell the keys and helps Maia in the passenger side. Terrell roars away, and some of the gravel hits the front window. *News sure travels fast around here.*

Maia's desk is empty, but sitting at my desk is Bryan. Of all people, Bryan. "Hey lady, you're walking. Look at your teeth!" *That southern accent*. I smile at him. He doesn't look very much changed yet, but I know the changes are coming for him, too. I sit down in Maia's chair and keep smiling. "How are you, Bryan? Have you been able to do any trucking yet?" He grins and nods his head, his beard moving with his cheeks. "Oh yes, ma'am. I've been all over the place. Love running that rig." He motions outside, and I can see his red truck with a long, shiny white trailer hooked up to the back. "Cook has been fixing food for me so I can just reheat it in the microwave and go. I sleep in my truck, so I keep my expenses low." "I'm so happy for you, Bryan. Where are you headed next?" I ask. Before Bryan can answer, I hear that voice again, the one that tingles in all the right places. "Wyoming." I stammer, "Oh… Okay…"

Bryan smiles wide, "Malcolm is going to ride with me on this trip. He has some business out that way." I look up at Malcolm with wide eyes. His attention is at the front of the office.

Standing in the doorway is a tall man with gray hair and Unibomber-style glasses, wearing a brown uniform. He is holding a clipboard. "You're all loaded, Bryan. Safe travels. I sure am glad we have you now. I really disliked loading out at night."

He nods, "Malcolm." Malcolm nods back, "Gerald."

Bryan slaps his hands on his knees, "Best load up, Malcolm. We have a lot of miles to cover, and those folks are going to be expecting us."

Malcolm nods and turns to me, his face softening, "Be safe while I'm gone, Ellie."

I tilt my head, "What?" But he is already gone, and I can see him walking around the front of the truck. Bryan says, "He sure moves fast, don't he?" I nod and stand to move over to my own chair.

I'm gonna get this computer going. Let's see what's been happening with the scheduling. I get into the system to find that everything is ok. The little bit I did the other day on the tablet wasn't even needed. I look and check the user information. Maia. She has been keeping everything status quo. I'm going to have to thank her for this, too. It even looks like she has brought in the rest of the warehousing crew. *I bet Chef is beside himself now.*

My original plan was to check on the vacation time. I know now that sick days don't matter, nor personal time. Nobody cares. There's a "policy," I know, but no one seems to follow it. *Everybody wants to work here.* It looks like vacation time has piled up for everyone. Other than the company-sanctioned breaks, nobody has taken any time off. *I know they have families and places to go…* It looks like Gerald has the most time saved up. I'm going to go visit with him. I stand up and walk through the office.

Gerald meets me at the door to the warehouse. "I know what you want, and I'm not going to do it." I'm stunned. How did he even know I was coming? *How did Terrell know you were trying to walk a dark hallway by yourself?* "Hi Gerald, I'm Ellie." He cuts me off, "I know who you are, and I'm not going to do it. I wanna work." He's quite gruff. "Gerald, I'm not trying to make you take off work. I was wondering if you didn't want to use your vacation time if you wanted to sell it back to the company. Like getting a bonus instead." His eyebrows raise, "I'm listening." I smile and ask him, "Could we sit in here?"

I gesture to the dining room. He nods. We walk into the dining room and sit. "Gerald, I am supposed to be working through everyone on the schedule. That includes all the folks in all the locations. You're my guinea pig. You're the first person I'm talking to. Not only because I know who you are now, but also because you have the most vacation time saved up from everybody." Gerald is staring at me. "Gerald? Are you ok?" He looks down at his lap. "Ellie, I've been gruff and standoffish for so long, I don't know how to be anything else. Most people just leave me alone. You're the first person besides Nathan to come talk to me like a human being. Thank you."

He is twisting his hands together. "Can I tell you something else, Ellie?" I smile a little, "Sure thing, Gerald, I'm all ears." Gerald starts, "I miss my family." I start, "Oh Gerald," and go to set my hand over his. He pulls his hand away. "I don't need your pity. I lost them in a wildfire many moons ago when we lived in California. Maia's situation just tears me up. I hate that she has lost some of her family now, too." I nod in agreement and just let the man talk. *Most people just want to be heard.*

"Do you think if I sell back my vacation time, Nathan would get upset if I donated it to the Quapaw Nation for rebuilding? I don't need the money for me, but I bet they are going to need to build new homes and buy necessities for anyone displaced." Gerald continues to look at his lap.

Nathan walks in behind me, "Gerald, that money will be heaven-sent. I will match the amount, and we can send it together with a truckload of building materials as soon as Bryan and Malcolm get back." Now I'm crying. *These people are amazing.*

Nathan turns to me, "Ellie, figure up his hours. Gerald, did you want to keep any vacation time on hand?" Gerald says, "Maybe a day's worth, just in case." All 3 of us disperse.

###

Nathan follows me to my desk. "He's a good man, Ellie, and he's been closed off for far too long. Thank you for talking to him." I'm at a loss for words. He continues, "Since Maia is gone, could you help me track down the building materials to fill the trailer? We will need to send bunks of 2x4s, drywall...." He lists off the items faster than I can write them all down. Nathan takes a deep breath, "Then Bryan can head back out with a trailer full of insulation." I nod and turn back to the computer, still dumbfounded. I google lumberyards and start calling. I also tell them that we are gathering common materials to send to the Quapaw Nation and every last one of them donates additional items to send along—many including fasteners and tools. By the time I'm finished, there are 3 truckloads worth of materials being delivered here so Bryan can take it all out to Oklahoma.

It's past 4:30, and I'm getting very tired. I need to figure up Gerald's time, and it's just beyond me. *Tomorrow....* Once again, I'm the only one here and not sure how to get back around to the "good" hallway. I walk to the dining room to find Nova sitting on the counter, wrapped around Cook. I smile, but I don't leave quickly enough. They catch me and, at the same time, yell, "MILDRED!" I laugh and turn back around. "Guys, I need to get back to the apartment, but don't know how to take the better hallway instead of the dark one." Nova hops down and chirps, "I'll show you."

Nova leads me past the back row of sales offices and Nathan and Mr. Louis' offices to a plain black door. She opens it to reveal a light-filled hallway that curves up and to the left. "Oh, thank you, Nova," I reply, and she is gone. *Back to the make-out session...*

I stop for a moment. *I haven't seen Mr. Louis in days...* I turn back to his office door and knock gently. I don't hear a response, but I hear something like a bear growling. Mr. Louis is asleep in the chair in front of his fireplace. His head is all the way back, and he is sawing logs. Grinning, I gently shut the door and go back to the hallway door. Over my head is a bank of skylights; I can see the evening glow shining through them. I keep walking, but the further I walk, the skylights disappear, and I can tell I'm back inside the rock face. I keep walking to another set of double glass doors, but one says Apartments across the front. *Well, duh....*

Which one is Malcolm's? It wouldn't be right of me to start opening doors, so I resist the temptation to snoop. I didn't realize how many apartments are here and the hallway is longer than I thought. There are 12 on each side—a*lmost the whole alphabet.* I keep walking down to B.

###

After all the walking today, I'm super thankful for these ballet flats. I kick them off right inside the door. Mr. Mittens greets me in the kitchen, sitting upright on the counter and waiting for his supper. *Yeah, buddy, I'm hungry too.* I'm not in the mood for much, so it's fried eggs again. I set a little bit of the yolk aside for Mr. Mittens, and he devours it, licking his chops like he just caught the biggest mouse. I giggle and, in my baby talk voice, "You're such a big bad hunter, my wittle baby boy." I'm scratching his ears. Then I hear, "I am?" coming from outside my door. Mr. Mittens takes off. It's Nathan.

"Come in, Nathan!" I yell over my shoulder. He steps in, and while there is a smile on his face, it falls at the sight of the roses. "Where… where did those come from?" he stammers. I'm looking at his face, and you can tell he is crushed by this. I don't know whether to tell him the truth or smooth it over. *The truth wins, even when it hurts.* "Nathan, I…"

He cuts me off, "Ellie, I don't want to know." And he waves his hand in front of his face. "Anyway…" He recovers himself, "I wanted to bring you this." And he sets down a glass award in front of me. "What's this, Nathan?" I pick it up; it's very heavy. "Dad dug it out of the cabinet in his office. It's a Good Samaritan award that he used to give out when employees would do kind things for each other. He felt you were deserving of it. Next time somebody does something nice or really cool for someone else, pass it to them." I hand it back to him. He looks startled and hurt. "Ellie, don't you want the award?"

I laugh and shake my head. "No, you goof. You snuck in and charged my phone in the night so I was up and going today. If you hadn't done that, I wouldn't have even known if it was night or day or even what time it was. I wanted to thank you all day but didn't see you except when we were working with Gerald." He sets the award down on the counter and wipes his face with his hands, then up over his hair like he is flabbergasted. "Ellie, I don't know where to go with this. With us."

I stare at him. I do not want to have this conversation right now. *But here we are...* "Nathan. I need to share a story with you. Would you come in and have a seat?" I gesture to the couch, and he goes and sits. He has a deer-in-the-headlights look on his face.

I stand in front of him, "I need to tell you why I'm single." His mouth forms the word "Oh," but no sound comes out. I sit down next to him. "I've been living alone, well, with Mr. Mittens, for almost 6 years now. I followed a boyfriend from college all the way to Chicago. He was going to be a lawyer, and I was going to be an Executive Assistant somewhere. We had finished school and moved in together into a dinky apartment. I was young and dumb. It never once occurred to me that we wouldn't get married. It never occurred to me that he would take all our money, go out with other women constantly, and berate me to the point of tears when I pointed out his shortcomings. There were other instances, but things came to a head one November evening, and he was so angry he grabbed my arm, flung me down in the chair, and stood over me so I couldn't get up or get away. He lorded over me, spitting in my face as he yelled how stupid and crazy I was. No one would believe me; they would all believe him, and he was going to be made partner at work. I was just a secretary and would never amount to anything."

I stop to take a breath and check Nathan's face; it's passive. "What he didn't know was that I had started a separate savings account through my job that deposited part of my paychecks prior to the rest being put in our joint account. Every time I got a raise, I added the raise to the savings account. I told him I got demoted, which gave him more ammunition to spit at me every day, but it gave me a nest egg to rely on when I was able to leave. I didn't know when it would be until that November night." Nathan's eyebrows raised.

"When I was pinned under him in that chair, fearing for my life and my sanity; *in that order*, I decided I was finally going to stand up for myself. He was going to be done with me one way or another. I screamed, and screamed, at the top of my lungs for help. You know how those apartment buildings are, everybody can hear everything. The front door banged open; the big guy from the apartment above us had broken in the front door. He lifted Brett off me. Brett was so shocked and continued yelling obscenities at me while the guy pulled his arms behind his back and hog-tied him with the string from a trash bag. The big guy dropped Brett onto the floor from shoulder height. Then he dialed 911 on his own cell phone, told them the address, and left. I was too stunned to move. And thankful."

I stopped to check Nathan's face again. Eyebrows raised, very quiet, eyes wide. "The sheriff's department arrived and arrested Brett. They took my statement and gave me the address of a women's shelter. I asked about the guy, and they said they tracked the burner phone to a trash can outside the building. When he saved me that night, he had to uproot his life from whatever he had been doing." Since Brett was in jail and my front door was severely damaged, I took the time to pack up and leave the apartment. Thankfully, I had left my car in my own name. I took the checkbook and the credit cards out of his wallet and drove to the nearest hotel. The next morning, I went to the bank and drained the rest of the checking account. I left him $5." I stop and grin at Nathan. All he says is, "Wow."

"God smiled on me that day. A buddy from college called out of the blue. He was getting divorced and didn't know if I knew anyone needing a house. He had to unload it fast before it became a point of contention in the divorce. I hadn't talked to him in over a year and he calls with this solution out of nowhere. I asked how much. He said the $109,000 he paid for it. He had already moved out. I could move in the next day. Within a week, I had my down payment deposited into the bank; and an approved mortgage. We closed on it that Friday. Then, I found a different job closer to home. Mr. Mittens was living under the shed behind the house, so I brought him inside. I didn't press charges on Brett. He wouldn't need a domestic battery conviction to be made partner."

Nathan coughed. "And you've just been the same old Ellie every day since then?" That got my attention. "Yeah, Nathan, I have. I have no doubt that a relationship would do me some good, but I've not been willing to entertain dating again. That fear is real, and it was a traumatic experience all the way around." Nathan interrupts me, "And not a relationship with me either; we are too volatile around each other." I sigh, "Yes, that scares me a lot. I've not been mad enough to beat on someone's chest in years." I laugh a sad little laugh.

He interrupts me again, "But you would be good with my brother. He seems to calm you… or at least being around him seems to shut you up." *Busted. The truth hurts, remember? Yeah yeah…* "You're right, Nathan, I am attracted to Malcolm. I have no idea how to handle it either." Nathan stands and holds his arms out for a hug. I walk into them. "The roses are from Malcolm, aren't they?" His muffled voice vibrates through the top of my head. "Yes, Nathan, they are." "That's good, that's good." Is all he says. He turns to walk away, slips out the door, and goes into the hallway. He left the award behind.

***

CHAPTER 12:

Things settled into a routine after that. I got up at my usual time, and since my "commute" shortened so much, I was the first one in the dining room each morning. Cook was usually ready with a plate for me with each day's special. I would then help him set up the coffee bar before going to my desk. I managed to work through 5 groups of employees in Volley. Most everyone chose to sell back their vacation time. All of them citing that the company-sanctioned retreats were enough downtime for themselves and their families. I was also the whole show in the office with Maia gone. Caleb took over Terrell's duties while he was away. Bryan and Malcolm returned on the third day. Immediately, Bryan and Gerald loaded the trailer full of building materials, and he took off for Oklahoma. Malcolm just ignored me. Flat out, he turned his head away from even looking at me if I was in the same room with him. It hurt. *Whatever, though, it's not like you were attached.* I ate lunch alone at the counter in the dining room most days, sometimes sneaking away with just a sandwich to my desk. I didn't know most of the other people from the warehouse and harvesting crews yet. With Malcolm acting the way he was, my confidence was pretty low. Maia and Terrell were gone for a solid month. (Maia still hasn't returned to the office.) It took Bryan 4 trips that week to get all the supplies delivered to the reservation. *But I haven't seen Mr. Louis.*

I am all healed up now. I'm moving fine, completely released from the doctor's care. And I'm tired of the 4 walls in the apartment and the 4 walls in the office. I'm tired of wearing the same clothes over and over and trying to make them look different. I want to go home.

I look up the phone extension for the garage and Monkey answers on the 4th ring. I can hear heavy metal music screaming through the earpiece. I yell, "Hey, Monkey, if you can hear me, I need my car. This is Ellie." And I hung up. Not 2 minutes later, Monkey is at the front door and tosses me the keys. "Boss man isn't gonna like this, Ellie." *That's strange.* "Monkey, it's my car. Why is it going to upset Nathan if I use my own car?" He shrugs and turns around, mumbling something about, "Your funeral…"

By 4:45pm, there is nothing going on in the office; the phone hasn't rung for a couple of hours now. I have gotten quite a bit of work done today, so I shut everything down. I start walking back to the hallway that goes to my apartment. I don't know how to gather my things and get them to my car efficiently, but I can take the necessities and the cat and head home. I will clean out the apartment when I can find someone to lift and haul the rest. I get to the apartment door and hear something going on inside. It sounds like an Italian man…. Singing…

I open the door to find Malcolm holding Mr. Mittens up in the air, singing very loudly to him in Italian. Mr. Mittens loves the attention and is limp in Malcolm's grasp. *Traitor.* Garlic, tomatoes, and basil all assault my nose. I rap on the open door, very loudly.

Malcolm stops mid-swing with the cat, turning his head to look at me. Mr. Mittens is wearing a scowl that says I just interrupted the best thing that has ever happened to him. Malcolm's face is red. And I mean as red as the tomato sauce simmering on the stove. He sets the cat down.

I shut the door, and we just look at each other for a long minute. He shakes his head and walks towards me like a lion stalking his prey. He presses me up against the wall; a picture falls down, and the glass breaks. We don't notice it. The kiss is searing, long, languid, sucking each other's tongues. We turn our heads back and forth, investigating each other's mouths, bumping teeth, biting lips. We're dancing and kissing like it's something we've done a thousand times. *What was it that I wanted to do to him? Oh yeah.... THAT. Down girl, down....*

I push him off me, and he pulls away. We are both panting. I look up at his face. Those hazel eyes are burning hot. *OH... He wants that, too. Okay, we gotta cool it here.* We both start to talk all at once. "Malcolm..." "Ellie...." I giggle, he chuckles, and he motions to the bartop. "Sit, please." I sit as I am told, and he fixes me a heaping plate of spaghetti with meat sauce and then sets a big hunk of garlic bread on the side. "I didn't make any vegetables." He grins and adds, "Cook can worry about vegetables tomorrow." Malcolm fixes a plate for himself and sits at the bar next to me.

He motions to the roses. "Looks like those have seen better days." I nod and finish chewing my bite of food. "I haven't had the heart to throw them out; I didn't know if I would ever get flowers again." He sees the tear trickle down my cheek, and his face softens. "Ellie, we need to talk. Right now though, could we finish our food and get to know each other better?" I agree and go back to eating.

Then I ask, "Which Nonna taught you to make all this?" He grins and goes into a long explanation of the tight-knit families over there and how they welcomed him with open arms. I nod. "That sounds like a wonderful experience." And we fall silent. Me, knowing I would never have a family situation like that, and him knowing he had told me something that upset me; again. I interrupt the silence, "Please tell me more. I don't mean to get introspective. Your voice makes me tingle." *Oops, I said the quiet part out loud there...*

"Tingle, eh?" and he reaches over to tickle my side. *Oh no, no tickling.* "No! No! Stop! No tickling; I will throw up my dinner!" He stops, but he's still laughing. I'm thankful he listened to me. "Explain to me this tingling you are feeling then." And he scoots his stool up next to mine. *Okay, that's enough.*

"Malcolm, we need to talk big and talk big right now." I use my stern, no-nonsense voice. The smile wipes from his face in an instant. Mr. Mittens even sits up and moves to the bedroom. Malcolm walks over to the couch and sits down in the exact same spot Nathan sat in the other night. *Not all this again...* I ask, "Have you talked to Nathan about any of this?"

Malcolm hangs his head. "Yes, I have. I have his blessing. I've tried to keep my distance from you out of respect for you both. But it looked like I was hurting you by doing that. I finally decided just to let you see who I actually am." My temper starts to flare.

"First, I want to thank you for supper; it was very good. Second, I would like for you to leave my apartment." Malcolm replies, "Oh, ok. I didn't mean to upset you, Ellie." He gathers his jacket and steps to the door. He looks back at me. *If he doesn't hurry up, I'm gonna lose it.* "Thank you, Malcolm. I will see you tomorrow in the office." He shuts the door gently, and I hear the latch click.

I sit down on the couch and start to cry. And I think all the hurt from 6 years ago pours out, too—no sense trying to drive home now. My eyes are so swollen I can't see very well. Everything is blurry. And that makes me cry even more because, really, *I shouldn't even be alive right now.*

###

The next morning, Cook looks at my face. "Ellie, are you ok?" I look up from pushing the food around my plate. "Man trouble." And I look back down at the plate. Cook slaps the counter with his towel. It was such a loud CRACK that I startle and fall off the stool. He runs over to catch me and helps me up. He holds onto my shoulders and keeps me at arm's length. "Now, Ellie, you listen up right now." I look up at him, and I've already started to cry. Big fat tears are rolling down my cheeks. *Not again.*

"Cook, what now? I'm not in very good shape today." He softens, sets me down on the stool, and sits across from me on the next one. "You did me a service by telling Nova she needed to tame herself." I look up in astonishment. "I didn't tell her that Cook. I thought it but didn't say it out loud." My brain takes off running with this information. *Oh. Holy. Shit.*

Cook continues, "She told me you told her to calm down and let me be me." I'm still back at *holy shit; she heard my thoughts.* That means that everyone has heard my thoughts, the snarky and the sad and the lonely… *Oh, God in heaven.* I put my face in my hands and sob. Cook is rubbing my back with a solemn expression on his face.

Malcolm walks in and roars. "YOU TOLD HER?!" I look up at the murderous expression on his face. He is quickly crossing the room. Cook stands up, "She figured it out for herself." He puts himself between me and Malcolm. And, of course, here comes Nathan. "What is the problem here?" He booms. Nathan looks from me to Malcolm and then to Cook.

Nathan walks up to Cook, "Thank you for this; it will be easier now." Cook stands down, walks to the kitchen, and glances back at me. I nod back to him to let him know I will be ok. Nathan takes Cook's spot in front of me and squares up to Malcolm. Malcolm starts forward, and Nathan puts his arm out. "Stop Malcolm. She is already upset; let's not make it worse for her."

All of a sudden, in my mind's eye, I see a flash of the peach-colored roses. I hear a faint whisper like it's only in my mind; "*Whatever she needs. Everything for her.*" I tilt my head, tap Nathan on the shoulder, and look up at him as he turns around. I try to use my mind to ask him, "Is it reciprocal?" I hear back the faintest whisper, "*Yes*." Malcolm looks at me, barreling his focus into my eyes.

I stand up, "Nathan, I'm afraid I'm going to need to take the day off." And I stride out of the dining room, walk up to the staging area, and swipe a set of keys. I take off in an ATV, driving to the left instead of the right, driving as fast as I can. *This would have made more sense in the car. Stop that. You're vulnerable, too.* Now I realize my thoughts haven't been safe all this time. I had no idea. I think, as I race through the corridors. There were signs, definitely signs that I missed or shrugged off.

***

CHAPTER 13:

I'm a mess, the ATV is a mess, and nothing is ever a mess around here. I slow down to look around. I'm in completely unfamiliar territory, and there is mud—quite a bit of mud, mud that looks deep. I don't want to step out of this ATV. *At least I don't have to worry about getting stuck.* I also realize the path has narrowed, and it is definitely now more of a path than a road. I'm creeping slowly at this point. There is no place to turn around. Only through. I press forward, but I'm getting more freaked out by the moment.

I realize there are no lights on the ceilings, only on the ATV. I don't have my phone or a radio or even a hard hat. *And I don't have any clue where I am.* It's colder in this part of the cave as well. Finally, I just stop and sit in the ATV. The mud is above the tops of the rims, and I'm afraid of going deeper into more muck. I'm cold and getting hungry. *If ya'll can hear my thoughts, this is what it looks like where I am. I can't turn around, and I'm scared.*

Then, the ATV sputters and dies. *Out of gas. Fuck.* The headlights are still on, though. *Let them burn.* I don't see anything ahead of me. I had thought the cave was quiet before. It sure isn't. I can hear air rushing past my ears. Something is scruffling along the edge of the pathway and there's the occasional FLURP of a water drop falling from the ceiling. I don't know how long I can sit here. The headlights dim and finally go out. I'm on high alert inside my mind, but my body is cold and tired. I start to drift off. It seems like hours are passing, and there is no sign of anyone or anything.

###

I'm shaken awake. I scream as I face a very dirty man in a hard hat with a headlamp strapped to it. He is wearing overalls that are caked with mud. "Ellie! Ellie girl!" the man is yelling while shaking me. I stand up and fall back into the passenger seat. I'm in full freak-out mode. I don't want to step down into the mud, but I will if I have to. Finally, the man digs his phone out of his pocket and turns on the flashlight. He shines the light onto his face. "Ellie, it's me." I calm enough to see it's Mr. Louis, as filthy dirty as I've ever seen him.

He points to the driver's seat, "May I sit with you?" I nod, and he sits down with a sigh. "Oh, it feels good to sit in a padded seat." I've come to enough to finally ask, "What are you doing down here, over here, wherever here that we are?" Mr. Louis chuckles. "I never can get fully away from doing some of the manual labor out here. You know I started out as a farmer?" I nod, and he continues, "Once I discovered the mineral, I also found this cave and have been working at exploring and mapping it for about 80 years now. When I fainted awhile back, it really opened my eyes to the fact that my time might be more limited than I think. I'd been putting off getting back out here to explore. I do enjoy my creature comforts." He pats his belly for emphasis. I nod, but I don't speak because I know I'm going to start shivering. "Oh, Ellie, I'm so sorry. We all tend to lose our manners around you. You just make everyone feel comfortable.

Give me a moment, and I'll be right back." He hands me the phone with the flashlight still on. It's unlocked, and I hurry up and shoot a text to Nathan, "This is Ellie. I'm with your Dad. If you know where he is, could you come get me? The ATV died." *I'm sure there's no service down here anyway...*

Mr. Louis comes sloshing back with a heavy farm coat. *Haven't seen or worn one of those in a while, have ya?* I slip my arms into it and instantly feel warmer. It smells of dampness, but I do not care. "Thank you, Mr. Louis." I squeak out between gritted teeth. I hand the phone back to him. Mr. Louis starts, "Ellie, why are you all the way back here? How did you know where to find me?" I talk through gritted teeth, "I didn't, Mr. Louis; I was extremely upset and took one of the ATVs for a ride to clear my head. Now I'm stuck because it has died. Both the gas and the battery have given out." Mr. Louis sticks out his arm, "Come here and sit up next to me to warm up. The boys will be here soon." I look at him in the dim light from the phone. "How do you know, Mr. Louis?" He points to his head, "I can hear them. They are pissed; at each other, at you, at me." He chuckles, "They are just plain pissed!"

It's sure taking a long time for the guys to get here. Mr. Louis keeps me awake by visiting about the old-days. He is reminiscing about working as a farmer side-by-side with his wife. I don't broach the subject of his wife; I just let him talk. He finally asks, "Would you like to see what I found today?" He reaches into the top pocket of his overalls and pulls out a few small stones. "They look blue to me," I say. His smile lights up the whole area. "Sapphires. The first found in Illinois."

I can faintly hear an engine with my ears. But in my mind, I hear a loud barrage of conversation between Nathan and Malcolm. Everything from "*She's mine*" to "*You created this whole mess*". Mr. Louis is chuckling. *How did he hear them so soon? Probably because he knows them both inside and out. They are his sons, duh...* I hold my hands over my ears to try and curb the noise of the two of them fighting. But as they pull up, all I can hear outside is the engine. They aren't speaking to each other, their expressions angry. The arguing I'm hearing between the two of them is completely non-verbal. *How do I turn this off? This is terrible!* Nathan turns off the ATV and steps down into the muck.

I look down and see he took the time to put on those army green muck boots. Malcolm is wearing a pair, too. *Somebody is going to have to lift me up and set me in the other ATV.* Malcolm says out loud, "Don't mind if I do!" He picks me up, slings me over his shoulder like a sack of potatoes, and deposits me in the passenger seat of the working ATV. He also clicks the seatbelt over me and plants a light kiss on my forehead. He levels his gaze with mine, pulls on the strap of the seatbelt, and says, "So you can't get away again." Then he reaches into the bed and pulls out a gas can and battery cables.

Nathan is talking to his Dad, "You ready to get out of here, Pop? You've been back here for three days, and you have to be low on food." Mr. Louis nods and trudges back through the muck to grab his tools and supplies.

Nathan is shaking his head. "That man is so stubborn." I yell, "the apples didn't fall far from the tree!" And Malcolm flicks a tiny speck of mud directly onto my nose. Accosted, I reach down the side of the machine and get a whole handful of mud. I'm ready for when he walks by again. But he doesn't come this way. *Oh, that's right, he already knows what I'm planning. And he can hear that I'm reminding myself of that fact. Sigh.....* Now my hand is cold, wet and very dirty. I've ruined my nails. *Brilliant.*

Malcolm chuckles and turns to Nathan, "Are you listening to her? She's funny." Nathan shakes his head, "No, I'm listening for Dad. He hasn't come back with his supplies yet." I look up, "He didn't find me for a long time. I was lonely and cold before he got here." Nathan looks indecisive. Malcolm stops him, "Give him a few minutes, brother; we don't know how far back he was mining."

After awhile, we hear the slosh, slosh, slosh of a man traipsing through a mud pit. Mr. Louis arrives out of breath, his arms loaded with a shovel, pick, cooler, and thermos. "I.." He takes a breath, "I didn't realize how far back I had gone. If I hadn't heard Ellie earlier, I would still be back there mining." He grins wide, "Wanna see what I found, son?" He lifts his hand to reach into the top pocket of his coveralls. Nathan rolls his eyes, "What now, Dad? An arrowhead again?" Mr. Louis looks hurt. Malcolm steps forward, "What did you find, Dad? Something new?" Mr. Louis grins and grabs the little stones out of his pocket. "Sapphires, the first in Illinois!" Malcolm's eyes are wide. "That's amazing, Pop."

Nathan shakes off his skepticism and smiles at Mr. Louis. "Can you help us jump this ATV, Dad? Ellie did a number on it." He looks back at me and grins. Now I'm rolling my eyes. "Just because nobody told me about all this telepathic stuff and I had to find out the hard way and panicked." Malcolm chimes in, "Yeah, yeah. When you get better at it, you will find it's quite efficient." *Better at it? Hmmm....*

The guys go through the motions of filling the gas tank on the other ATV and hooking up the battery cables from one to the other. I don't know who yelled at me, but it startles me. "Hey, Ellie. ELLIE!! Crank the engine on that one!" I can't reach it with this seatbelt on, so I unclip it and reach over to turn the key. It starts up, and the guys are finally able to get the other ATV to turn over. *Whew.* I can hear Nathan "telling" Malcolm to take off with me on this ATV. He is going to ride with his Dad. *Oh no. I'm not going to go through another round of last night's fiasco.*

I stand and step down, intending to put myself in the bed of the other ATV. And my foot goes squish… all the way down to the rock floor. The mud is cold and wet; I yelp at the shock of it. Malcolm is laughing full-on at this point. "Go ahead, smarty pants, pull your foot out of the muck." I grimace and pull my foot up and realize that the mud kept my shoe. *Oh, this is just great; "smarty pants" is accurate.* Nathan and Mr. Louis both chuckle at that one. I sit my fuming rear back into the seat and re-buckle the seatbelt.

Malcolm puts his arm around the back of my seat to look back as he reverses the ATV. I can smell his deep masculine aroma, along with whatever soap and aftershave he uses. *Lord, help me.* Malcolm drives us backward for quite a ways until we get to the outside of the harvesting chamber. He is able to back into a little turnaround and then speeds forward across the main hall to the drive that takes us to the apartments. He helps me out of the ATV and stops at my door marked "B." I don't know if anyone else saw us, but I'm sure I looked a sight in Mr. Louis' coat, one shoe, one muddy foot, one muddy hand, and a spot of mud on my nose. *Shit. I don't have any more shoes.*

###

Malcolm opens the door to the apartment. Mr. Mittens meets us there with a loud meow and he rubs on Malcolm's leg. *You're still a traitor.* Malcolm is grinning at me. He helps me out of the coat, clarifies that I'm ok, and steps out of the apartment, shutting the door softly. I sigh. *Right now would be a really good time to try that bathtub.* I decide it is best to just strip off everything here in front of the door instead of hauling the muddy everything through the apartment. I turn to head back to the bedroom and a hot bath. Right before I step into the bedroom, I hear Mr. Mittens meow again. I turn back around just in time to see Malcolm walk right in the front door with another bouquet of roses. He stops in his tracks and stares. *Well, he's seen it all now.* All I have left on is my underwear. I stomp forward, and I'm pissed. All my parts jiggle at every stomp, and I don't even care. MALCOLM LAZARUS LOUIS! Do you even know how to knock?!" He looks dumbfounded, sets the roses on the counter, and backs out the door. *Damn, those roses are beautiful.* I decide to deal with them and him later. *I'm going to go take this bath.*

CHAPTER 14:

And it is glorious. As much hot water as I want, jets working on all the right spots. I even try out some bubble bath that's been sitting next to the tub all this time. *Why didn't I use this thing sooner?* I'm squeaky clean and smell like a bucket of flowers. And then I realize I'm entirely out of clean clothes. *I've got to go home.* I pull the best-smelling pair of yoga pants and a t-shirt out of the dirty laundry pile. I slip on those infernal hospital slippers they made me wear in the infirmary and grab my purse.

I pat Mr. Mittens on the head, "I'm running home, buddy. Are you missing anything?" He doesn't even give me the time of day. I walk around the good hallway to the office. My car is still sitting out front. I grab my keys off my desk, close the front door softly behind me and get in the car. I've missed driving.

###

I toodle on home and find a black SUV in *my* driveway. *Umm….* I use my key in the front door and startle a woman cooking at *my* stove. I hear loud footsteps coming up the basement stairs of *my* house. I'm still standing in the middle of the living room with my mouth hanging open. Dumbfounded. I look around and see the furniture is different. A tall man walks around the corner from the steps. With a stern expression on his face, he walks towards me and asks, "Can I help you?"

The lady wipes her hands on a kitchen towel and walks up next to the man. "Umm…" is all I can manage. Finally, I snap out of it and ask, "Why are you in my house?" The man straightens and says, "We rented this house a month ago. Talked to a guy named Nathan. You must be Ellie; he said he was your brother." *I'm gonna kill him. I don't care what kind of a mess it makes.*

Just then, Nathan steps inside the front door and nods to the couple, "Robert, Darlene. Good to see you." Nathan steps right up behind me and puts his hands on my shoulders. Robert looks up and motions to me. "Nathan, didn't she know?" Nathan cuts him off, "She was in the hospital; I haven't had a chance to talk to her." I can feel my face heating up. I know Nathan can hear my thoughts, and they are not friendly *at all.* I turn to him and ask if we can talk outside. We both step out the door.

As soon as the front door shuts, I scream and start beating him with my purse. Over and over, I hit him. He is crouching down and begging me to stop. The curtains move on the front of the house and I look up to see Robert and Darlene watching. Nathan stands and grabs ahold of my arm. He forces me into the Jaguar's passenger seat. He runs to the driver's side.

As soon as the door clicks on his side, I start in. "You son of a bitch! How in the hell can you do this to me?! If I kill you and it makes a mess, I am not cleaning it up!" Nathan is shrunk as far across the car from me as he can possibly get. "Where are all my things, Nathan?! NATHAN!!!" I don't know how many minutes it goes on, but I rant a little while longer. I don't think I've ever been this angry in my life. Finally, I can speak in a fairly normal tone and ask, "Do you just enjoy fucking up my life? Is that it?"

A loud, red, lifted, Dodge pickup hauls ass, right up into the front yard. It stops parallel to the Jaguar. Malcolm jumps out, "*Don't kill him, Ellie*!" I realize I heard him in my mind. He is standing outside the car on the driver's side. Nathan glances over at me and gingerly opens the car door. He steps out, and Malcolm gets in—almost in one fluid movement. Nathan gets in the still-running truck and backs down to the street. He roars away. *Robert and Darlene didn't know they were getting a show along with their lunch today...* I roll my eyes so hard that it's painful. Malcolm starts to speak but thinks better of it. A few minutes pass and we just stare at each other. Finally, he says, "Are you calm enough to drive?" I nod, and he allows me to get out of the Jaguar. Before I can get my last foot out, he stops me with a hand on my arm. "Could we go for a drive together?" I nod, and he says, "Meet me at the office and leave your car in its usual spot."

I pull up in front of the office. The red pickup is already there. Monkey is standing in front of the door, waiting for me with his hand out for the keys. I glare at him and place the keys in his hand. *At least the kid is smart enough to keep his mouth shut.* The Jaguar rolls in behind me, parks next to my car, and Malcolm hands the keys to Monkey. Malcolm motions to the truck and takes me around the passenger side. He opens the door for me and helps me lift my slippered foot onto the running board. *I haven't been in a truck like this since I dated that guy in high school.* I hear a growl next to me.

Malcolm was in the truck before I had time to settle in the seat. "What was the growl for?" Malcolm looks a little sheepish, "*You dated that guy in high school.*" I hear in my head. I reply out loud, "Oh, for Pete's sake. That was 20 years ago!" "Don't care," he replies. *My Word, this day....*

###

Malcolm backs the truck up, turns around, and exits the driveway away from the quarry. We drive for several minutes, both lost in our own thoughts, him listening to mine, but I cannot hear his. He slows down, and we turn into a rough driveway - it's a cattle guard with a blackboard fence on either side. *Hmmm... It's pretty nondescript.* We bump and rumble over a path through some pasture ground for longer than I imagine. It's not in great shape, definitely needs re-seeded and this is neither a driveway nor a lane, but it's worn like tires have been rolling over it recently. Wherever we are, it's the middle of nowhere. I can't even see a power line from here. The truck dips down like it's going down a hill and then levels back out. I do see some trees further ahead, but cannot tell what is coming up. There's a crane truck up ahead. *What in the?* There are also a couple of storage containers parked off to the side. Malcolm keeps driving, and we get into the shade of the trees, turn a little to the right, and in front of me is a brand spanking new 2 story farmhouse. *How? What? Well, it's gorgeous anyway...*

Malcolm stops the truck right in front of the garage door. There's some large gravel in what would be the driveway but it hasn't been finished yet. The house is all white with black trim. The siding is board and batten like a barn. The garage doors and front door are black. The roof is dark but doesn't look as dark as the trim. There's a deep front porch all across the front of the house.

I start, "Malcolm, what is this place?" But when I turn to look at him, he is gone. He has come to my side of the truck to help me. I start again, "Malcolm…" He puts that finger to my lips again. *Zing!* "It's not finished." *So?*

He leads me to the front door, opens it, and gestures for me to step inside. The inside is completely empty. The walls are studded out, but no drywall has been done. There's only a subfloor. There's no insulation in the walls, either. I turn to look at him. He's gone again. *Now what?* I look up and see a loft, and Malcolm is hanging over the stub wall, looking down at me. "What do you think of it?" I shrug. "It's beautiful, Malcolm, but what does it have to do with me?" *I can't afford this place.* There are a couple 5-gallon buckets sitting upside down in what I would guess is the kitchen. Malcolm gestures over to them, "Meet me over there."

I go over to the buckets and sit. Malcolm sits down next to me, close enough to touch. *Oh, the smell of this man…* He starts, "Ellie, I need to tell you a story. I'm not going to go into the details of the breakup part, but you need to know what happened. Ok?" I nod.

Malcolm takes a deep breath and begins, "I left a little over a year ago because my heart was broken. I met a girl outside our company and dated her for several months. She turned out to be someone none of us could trust. Even though I had deep feelings for her, she took them and crushed my heart. I had been making plans for us for the future. Not only didn't her history check out, neither did her mind." He points to the side of his head. "She had me completely fooled. I didn't even get to hint about what we do around here, so there was no breach of our secrecy. I was shattered when I found out what she had been hiding from me the entire time we were dating. She was able to hide her mind from me." He sighs.

"I wanted to be away from here and any public place we had ever been to. I also knew I would lose it if I saw her out and about. So I left and went to Italy. Our office and warehouse family over there welcomed me with open arms. There was plenty to do as it is still a small office. I was able to occupy my mind and be useful at the same time. They were pretty upset when I left several weeks ago. Dad explained Nathan's predicament to them, and they understood. I'm still in contact with them almost daily."

He looks up at me, "I wasn't happy to be returning home, even if Nathan did need me. I didn't want to face everything I had tried to emotionally bury. But that's part of being in a family that runs a business, and I came back home."

His eyes well up. "The *minute* I laid eyes on you, I knew I was in trouble. Nathan had already told me quite a bit about you, but I had no idea I would feel pulled to you as soon as we met. I knew Nathan was interested in you and didn't want him to miss his chance. You're allowed to make your own choice, even if it would kill me if you choose him."

He looks at me, "You doing ok there?" I guess I hadn't made a sound in either capacity of communication. I squeak out, "Ye.. Ye… Yeah."

Malcolm begins again, "I know I'm going to be back in this location for the rest of my life. I hate living in that apartment; it's too cramped, and there's no sunshine. I need to be close to home, but I sure don't want to live near anyone else I work with every day." All of a sudden, I realize that he understands. *This man understands why I'm so upset about my house.*

I look deep into his eyes, and he smiles. "Yes, I do understand. I have been kicking around plans in my head for a farmhouse for years now. I had never taken the steps to work on it because I didn't have anyone to share it with." *I thought he wanted to be alone.*

Malcolm chuckles, "Not alone, Ellie, just not around everyone at work every minute of every day. Once I returned and met you, I decided it was time to build the house. Granted, I wanted many features that normal people wouldn't need, but it was simple enough to integrate them." He winks. *Whatever that means.* "Anyway…I wasn't going to bring you out here until you had agreed to marry me." I stand up so fast I knock over the bucket with a clatter. I open my mouth, but nothing comes out. Malcolm stands and grabs ahold of my limp hands. "I know we are nowhere close to that. Please take a breath."

He reaches back and resets the bucket so I can sit back down. "I will wait as long as forever if I have to, Ellie; there's no time limit on this. Please have a seat." I realize I am still standing and sit back down. I let out a big sigh. He looks over, "You still doing ok?"

I start, "No, Malcolm, I'm not. It's been a long and crazy day already. This is just too much to process all at once. I was overloaded before breakfast. I'm so upset about everything that I haven't even thought about lunch, and I think it's almost dinner time. I just wanted to go home, and now I can't go home, and I wanna see my cat." I burst into tears. *What. A. Day.*

Malcolm purses his lips. *"I always fuck this up."* I reply out loud, "No, you don't. I'm overloaded. IF we are going to choose to be together, we have to learn to roll with each other's limits. You have to admit the last several weeks of my life have been far more exciting than the past 6 years."

Malcolm laughs, "You're right. For someone with a routine ingrained in them, you sure blew that out of the water by working with us." I giggle a little, and then my stomach growls. Malcolm pulls me up off the bucket.

"Let's go take care of some things." He loads me back into the truck and stops at one of the storage containers. He gets out of the truck, flips the latches on the container, and pulls the heavy door open. From what I can see from inside the truck, it's all my things from the house. *That's information I didn't need today...* Malcolm grabs a plastic tote and lifts it into the back of the truck bed. He shuts the storage container, gets back in the truck, saying nothing. Then he drives back across the pasture, over the cattle guard, and back onto the highway. As we arrive back at the office, Malcolm walks around to my side to help me down and leads me to the front door.

"Head on back to the dining room. Cook is waiting for us." He steps back outside, lifts the tote from the truck, and carries it in. He sets it on my desk. "Go on, Ellie, Cook is waiting. I'm just going to leave this here for now. I will pick it up on my way through after supper." I turn and head to the dining room; I really am hungry.

###

Cook smiles wide. "Ellie! I didn't know if you would ever talk to me again!" I laugh, "I gotta keep nice with the food guy." I punch him on the shoulder. He places a hand on my shoulder and whispers, "Are you ok? You look very tired." *He is so thoughtful.* I smile weakly, "Yeah, Cook, bodily, I'm fine. Mentally and emotionally, I'm wrung out. You got any grub?" Cook laughs and turns back to the kitchen. He walks out with two plates of food piled high. *Always too much food.* Cook replies out loud, "Too much, Ellie? You always eat all of it." I chuckle. "You're just fine, Cook. Your food is always fantastic, and I can't stop eating it! Even if it is too much."

Supper is a full-on turkey dinner: stuffing, mashed potatoes, and gravy. The man even remembered cranberry sauce. "Hopefully, this will make you sleepy tonight, Ellie. You need your rest. You're not too far off from that injury a few weeks ago."

Where is Malcolm? I sit at the counter, and Cook places one of the plates in front of me. Malcolm walks in, beaming, "I heard my name?" I roll my eyes and gesture to the plate Cook is holding.

"You were right; Cook is waiting for us. Get up here and eat." I use my no-nonsense tone. Malcolm sits down and starts shoveling food in. Cook asks, "Ya'll need anything else? If not, I'm going to go on for the night. Just put your dishes in the sink."

I reply, "Cook, you're amazing as always. Go on, he…" I gesture with my fork at Malcolm, "Is pretty easy to keep in line." Malcolm laughs, Cook laughs, I groan. Cook leaves, and we are left alone again. We both finish our food and Malcolm holds out his hand for my plate. He takes them to the kitchen and grabs my hand. *Zing! Whatever.* We walk back around to the office, and he has to let go of my hand to carry the tote. I hold the hallway door open for him, and he steps through. We walk all the way down to apartment B, and I hold the door open for him again. He walks through the living room and sets the tote next to the end of the bed. He looks at me, kisses my forehead, and turns to leave. Mr. Mittens comes running in with a "meow-ow-ow," as if he didn't want to miss his favorite person. Malcolm chuckles, kisses the cat on the forehead, too, and leaves. I look at the cat, "You're worse than I am."

I bend down to open the tote. It's a bunch of my at-home clothes from the old dresser. Yoga pants, t-shirts, plain undies. At the bottom is a whole pile of socks and a pair of sneakers. Ok, back in business for a day. I strip off and run through the shower; nothing fancy, just one showerhead. I fall into bed, my hair still wrapped in the towel.

CHAPTER 15:

I wake up in the morning with my face smashed into the pillow, my hair half in and half out of the towel. It takes me a minute to realize who and where I am. I must have totally crashed out. Sitting up, I see Mr. Mittens at the foot of the bed, curled up in a ball with his paw over his face, like nothing phased him in the night.

I stand a bit wobbly today. *Yesterday was long and stressful, remember?* Healing from the brain bruising is going to take time, I know. I just want to be able to trust my own feet when I stand up. *You should have died.* Okay, enough of those thoughts today.

I go to the tote and fish out the clothes and sneakers I found last night. It's just plain yoga pants and a very old but not holey boy-band t-shirt. I grab a pair of socks and slip on the sneakers. *This is as good as it gets today, guys. It's the furthest thing from office casual I've ever seen.*

I go to the bathroom to straighten out this mess of hair. The night in the towel took its toll, and I've got a tangled wad with pieces sticking out all over. I spray on some conditioner and start to pull out sections from the mass. Once I get that solved, I pull it back into a ponytail. There is no use trying to keep my hair lying straight today. *It can do what it wants back there where I can't see it.* I stop to grab my purse and realize I haven't charged my phone. It is super dead again. *Whatever.*

With the charger added to my purse, I step out the door. I walk down the good hallway and hear that voice singing again, still in Italian. I smile wide and open the door where I can hear it the loudest. And it's the farthest door from mine at the end of the hall. Yep, it's Malcolm's apartment. The setup is the same, but it's decorated completely differently from mine. He is standing in the kitchen, a towel slung low across his hips. He stops singing mid-word and turns his head slowly to look at me. There's a piece of toast in his left hand, the butter knife full of butter in the other.

I giggle and say, "Good morning, handsome!" And go to step back out of the door. He drops his breakfast, and I hear the knife clatter to the floor. He grabs ahold of my t-shirt at the shoulder to hold me and wraps me in his arms in one fluid movement. I'm looking up at him; he kisses the end of my nose. His hazel eyes are bright and swimming. He crashes a kiss onto my lips. In just a few seconds, he lets me go. Completely. He just let go. I fall into the hall tree and land on my butt on the bench. *Thank goodness for that thing.*

He walks towards me with a shocked expression on his face. "Ellie, I'm sorry. Let me help you up." He grabs ahold of my hand and lifts me up, pretends to "dust" off my rear end, and I feel the smallest squeeze. *Uh huh... rascal.*

I take a step towards the door, "I'll…" He stops me with a hand on my shoulder. "Please, Ellie, stay for a minute. Let me give you the nickel tour. Then we can walk together to the office?" I nod and turn around. He points to the kitchen, which is identical to mine, with the cherry cabinets and black granite countertop. But the living room has dark hardwood floors and an entire wall of built-in cherry cabinets with blue backlighting. The fireplace is built right into the cabinetry. Next to the fireplace is a HUGE aquarium, also with blue backlighting. I don't hear it bubbling or any pumps running. *Must be a first-class type deal.* The furniture is all deep brown leather, and the walls are dark. I cannot tell what color the wall are in this low lighting. He leads me into the bedroom, and it's more of the same: very masculine. The bed is a dark cherry four-poster with a canopy—big rustic lamps on either of the nightstands. The walls appear to be navy blue. He points to the bathroom. It's set up the same way as mine, but the finishes are all dark wood and beautiful dark granite with veins of black and white going through it. My mouth forms a little o and I glance up at him. *We sure could get up to a lot of hi-jinx in these fancy bathrooms. Stop that!* Oh, Malcolm heard it, and he is grinning like a Cheshire cat. I cough and say, "I'll just meet you back in the kitchen." I almost run to the kitchen.

I see the butter knife and the mess it left on the floor. I bend to pick it up and put it in the sink. I find a bottle of cleaner under there and spray it on the mess. After grabbing a couple of paper towels, I wipe it up, and the greasy spot is gone. *Good stuff.* I look at the bottle; oh yeah, it's my favorite, but it's a different fragrance. *Cool.* Malcolm walks up behind me and clears his throat. He is wearing jeans, and those boots, and a flannel shirt. I tilt my head. *Ohhh... all we need is the cowboy hat.* He grins and points to the hall tree. There it is, too, a black cowboy hat. My knees are weak.

He starts, "I figure if you're having a casual day, then so am I." He grabs my hand and leads me out the front door. I look back and see that his apartment has no letter. It's the door I assumed was a janitor's closet. *Sneaky.* Malcolm replies, "Nathan's is across the hall. See?" He points, and there's a matching door with no letter. *Oh no.... I bet Nathan heard me giggling.*

Malcolm chuckles, "No, he didn't hear you giggling because he is already at the office. He did just hear you think that, though." I blush and ask, "How do you keep people out of your head then?" He smiles down at me as we walk through the glass door marked Apartments. "You will get used to the feel of the thoughts coming and going. It was actually pretty advanced when you asked Nathan if it was reciprocal. Most folks can't send a message directly to another person for months after their intuition kicks in."

I interrupt him, "Intuition, that's what you call it?" He looks down at me, "We didn't really have another name for it since it develops naturally, and the skills of using it increase over time." I look at him skeptically. "I would have called it telepathy." He interrupts me, "Yes, but if it were entirely telepathic, all our communication would be non-verbal. *Like this.*" And through his mind's eye, I can see a sunset over some mountains in some faraway place. It wasn't like watching a movie either; it was him showing me his experience, and I could feel the feelings he felt while watching the sunset. "Is that Wyoming?" I ask.

He nods. "I did a lot of soul-searching on that short trip to Wyoming. You see, Bryan and I originally met when we went to community college together on the Eastern side of Illinois. Bryan is like an extra son to my Dad; he spent a lot of time with our family during breaks. We talked all the way to Wyoming and back. Well, when we weren't sleeping, he would drive for a while, and I would drive for awhile." *So that's why Bryan wasn't phased by the company, the mineral, and how different everyone looks.* Malcolm looks down at me and nods.

###

I look up. We are standing at the black door that takes us into the office. He places his hand on the doorknob and sighs, "You ready for this?" "Ready for what?" I ask. And then I realize what as soon as he opens the door. Mr. Louis, Terrell, Caleb, Cook, and Nova are all standing right inside the door, all their faces with wide smiles, beaming at the pair of us still holding hands. I let go of Malcolm's hand. "*Too late now*," he says in my brain, grabbing back hold of my hand. I sigh. I use my mom voice, "Okay, everybody, nothing to see here. Move along, Move along…" All of them chuckle and disperse. I lead Malcolm to his little salesman's office. Tell him to "Sit." and leave him at his desk.

Terrell is in front of me. "No sense fighting it, Ellie; it was going to happen." I blow a raspberry at him. He chuckles and walks to his office at the front of the salesman's row.

I look over and see Maia at her desk. "MAIA!" I scream and wrap her up in a hug. She doesn't stand but pats my back and then leans back. I take a few steps back, grab my desk chair, and roll towards her. I sit and look at her. She is pale. Her hair lacks its usual shine, and her eyes look dull. She isn't wearing any makeup, either. I grab ahold of her hands. I don't know what I expected. *But it sure wasn't this.* She raises her head and nods. "I know it isn't me. We didn't grab any of the mineral before we left, and I'm starting almost completely over. At least this," she points to her head, "hasn't left me. It's not helping that I've been grieving so hard." And she hiccups, a tear rolling down her face. "Maia, I'm sorry. I'm so sorry. I'm sorry for the loss of your family. I'm sorry I didn't think to send mineral with Bryan."

She smiles and looks up at me then, "But look at you! You've been on the accelerated program!" I stammer, "Yeah… well…" *You should have been dead.* She stands up and bears over me. "You stop thinking like that right now. You were meant to be here, and you are, and that is that. None of this survivor's guilt shit. I'm not allowing myself to think that way, and I'm not going to allow you to think that way." She is almost yelling at this point.

Terrell walks out of his office. "Maia, honey." She whips around to him, pokes him in the chest and says, "I'm going to go get coffee." He holds both his hands up like she stuck him up with a gun and she stomps away. He turns to me. I notice he is paler too. Some of his braids have grown out a little. He puts his hand on his head. "Yeah, I need a whole new look." I give him a sad smile. "I'm assuming what you saw in Oklahoma was pretty bad." "Terrible," is all he replies. He sends me a short snippet with his mind. I can smell the fire through his vision. The row of houses reduced to ash, smoke still rising in places; dead bodies scorched and laying where they fell inside said houses. "Ellie, we got there fast. The authorities hadn't even had time to catalog everyone. With Maia being a direct descendant of Chief Strong Eagle, she was allowed into the wildfire area immediately. It was pure devastation."

He takes a breath like he is holding back tears. "She lost her entire family. She is the only descendant left from that line. Women don't hold office in the Quapaw Nation. The entire hierarchy is in disarray." I'm tearing up now too. "Oh Terrell, that means…" and my hand flies to my mouth. He interrupts me, "Maia has to find a Quapaw mate to keep the family line." Both of our faces show the grief of his statement. Maia returns with her coffee. She looks from me to Terrell. "Ah. You filled her in." Terrell shakes his head and goes back to his office. *Could we get a normal type day around here? YOU wanted variety….*

The phone rings, Maia answers it without her usual vigor. Terrell returns to his office and I return to my desk. All our hearts are heavy now. I work in Volley for a few hours.

Malcolm comes to get me. "I can hear your stomach from my office. Food. Now." I laugh and sit myself firmly into my chair. "*You just TOLD me to…*" He rolls his eyes. "Ellie, would you have time to join me for lunch? I'm famished, as I'm sure you must be too." *That's better. Gotta train 'em early.* I look over to see if Maia wants to join us, but I see her in Terrell's office with plates. I don't blame them. "*They've got quite a deal to go through.*" I send to Malcolm. He nods and holds out his hand for me.

###

We walk all the way over to Chef's dining room today. "Something different, mix up the atmosphere a bit," Malcolm says as he holds the chair out for me. Chef meets us at the table with his hands on his hips. "Look at you guys!" He's grinning from ear to ear. He leans down and whispers in my ear, "I made stirfry." *One of my favorites.* Chef asks, "Beef, Chicken or Shrimp?" *Is it bad to ask for all 3?* Chef laughs, "Nope, all 3 coming right up."

Chef sends a helper out with the plates piled full of meats, veggies, and rice. Malcolm looks at the "kid" and sticks out his hand to shake. "I've not met you yet." I interject, turning my focus to the guy, "I bet you're Grant. I just entered your information into the computer." He smiles and says, "Yes ma'am. I'm Monkey's brother, just a year older. He told me this is a great place to work." He turns to Malcolm and shakes his hand, "It's great to meet you sir." Malcolm half grins at Grant, "What crazy nickname would you like?" Grant grins back, showing messy teeth, "Just Grant is fine sir. I never have understood Tim's obsession with Monkey." He puts air quotes around Monkey. All 3 of us laugh. Grant excuses himself to get back to work. *Wait til he finds out what working here will do for his teeth.* Malcolm grins wide and shows all his teeth. We laugh and get to eating.

Grant stops by to pick up our plates and right behind him is Gerald. *He has on new glasses!* "Hey Ellie, you know that one vacation day I have left?" I can hear how nervous he is. He is looking between Malcolm and myself. Gerald carefully sinks into the nearest chair. Malcolm and I are both smiling at him. We can see a woman's face in Gerald's mind's eye. I decide I had better reply verbally, "Sure Gerald, I remember. What's up?" Gerald stammers, "I.... I... was wondering." He rubs his chin and looks down at the table. "I'm sorry, I've never asked for a day off before." Malcolm stops him with a hand up. "Say no more Gerald. Take Friday off and go. You don't need to use your vacation day for that." Gerald beams at Malcolm. "Are you sure? I haven't seen Nathan or Mr. Louis anywhere for days. I had no idea who to talk to." Malcolm nods. Gerald exclaims, "Oh, she is gonna love it!" And he stands up so quick he bumps the table hard. Plates rattle. "Oops!" He is pulling an old flip phone out of his pocket while he walks away.

I look over at Malcolm. "I've never seen a man that age, acting nervous like a teenage boy." Malcolm rolls his eyes and looks over at me, "Some of us are better at hiding it." *Oh really. This is information I will store for later.* I interrupt his thoughts; I can't hear them, but I can see on his face the wheels are turning. "Hey Malcolm." I interrupt his reverie. "Yeah, babe?" *Babe? Saving that for later too.* "I have two questions," I say. He looks at me like I should continue. "One, where are your dad and your brother? Two, when you say I'm going to get skilled at using my Intuition, what do you mean? Like just now, I know you were thinking something and I couldn't hear it."

He looks sheepish. "You really wanna know what I was thinking?" I nod. He shows me with his mind the scene of me stomping across my apartment towards him, almost naked, with all my parts jiggling. But I also feel the rush of arousal he felt at the sight of me, along with a lot of tenderness. "OH" Is all I can squeak out. I recover enough to ask, even though my face is flaming. "How do you control who can hear what?" He replies, "We can work on it over supper. Walk with me; I want to tell you about Dad and Nathan."

He leads me out to the patio. *You can sure feel fall coming....* Malcolm guides me along the stream bank, walking slowly and holding my hand. *I will never get over that zing!* He smiles, "Remember the sapphires? I thought Dad might have been wrong about what they were but he isn't. After this many years, Dad has connections with people all over the world in many different industries. He took one of the smallest ones to a stone cutter up in the city. They have been friends for decades, Dad had him make Mom's wedding ring; that's how far they go back. I jump a little, "That guy must be close to 100 by now!" Malcolm grins, "103, actually, still cutting gems. Delmer says the gems keep him young. Personally, I think Dad shares some of the mineral with him, but neither will tell."

I interrupt again, "Your Dad didn't have wedding rings made until he was 70?" Malcolm grins, "When Dad was 70, he didn't look 70. He met Mom, who was in her late twenties at that time and didn't tell her about the mineral until after they were married. I don't know what age he told her he was." My eyebrows are way up in my hairline. "Malcolm, how old ARE you?" He rolls his eyes, "Ellie,

I'm telling you the truth when I say this: I'm 45." *All these ages and the math of it are messing with my head.* "So your Mom was about FIFTY SIX when she had you?!" I say a little too loudly. He laughs, a big booming laugh. "And 58 when she had Nathan. He's the baby." *Harrumph, the baby.... "We are getting off track dear."* Malcolm sends. "I don't care; you can't hit me with this kind of information without clarifying a few things." I cross my arms. He reaches for my hand again, "I will tell you all about Mom another time. Let me continue about the sapphires, please." I nod.

"Delmer cut the first tiny sapphire yesterday. It was exemplary. The most perfect blue sapphire he says he has ever seen, very, very few inclusions. Delmer wants to work with more, but Dad has to find more. This morning, Delmer sent off the first cut stone to another gemologist in Belgium. Dad and Nathan have been down in the mine for almost 2 full days looking for more. Nathan needs a break anyway. Neither one of us wants Dad going back down there alone. You know why." I nod. "Yes, I would hate if he had another spell without anyone around." Malcolm nods grimly, "I'm afraid It would be catastrophic." I agree. *It's also best if Nathan keeps his distance for awhile.*

I realize then that we have walked the length of the creek all the way around to the side of the building near the entrance. Malcolm walks me in the front door, deposits me at my desk and walks back towards his dinky office. *I hope Mr. Louis does something about that soon*. I hear Malcolm chuckle. *We have all this mind stuff kinda solved, but what about the moving fast thing? "We will talk about that later too!"* I hear all this in my mind, coming from said dinky office.

I turn to my desk and start working in Volley when an email pops up with the red exclamation point for **Urgent**. I don't receive many emails unless there's a new employee. I open it. There's a new employee, all right; THREE HUNDRED of them… *How can we just add 300 employees overnight?* This time, I pay attention when I read it. The email is from Malcolm. "Ellie, we are setting up a new division in Oklahoma. Please add these folks in Volley." *Oh, ok... whatever. I'll just snap my fingers.* I can hear Malcolm laughing in his office.

It takes me the rest of the day to set up the location departments and start the employee lists. *It would be great if there was a way to scan the barcode on these folks' driver's licenses instead of having to key all this in.* I hear a yell from Malcolm's office. "That's a great idea. Let me email the developer." *We can talk to the developer of the software?!* I will have to keep the rest of the list of folks for tomorrow. It's past 6pm and I'm still keying. Maia and Terrell are gone; Caleb never showed today and Malcolm is still on the phone is his office.

###

I wander over to the dining room. Cook is just finishing up for the evening. Nova isn't there either. I wave at him in the back, "Hi Cook!" He slaps the towel over his shoulder and walks up to me. "Have you heard the good news?" I smile up at him, "I will take any and all good news you've got." He motions for us to sit. He stretches his legs out in front of him and crosses his ankles. His hands are folded on his belly. Cook is grinning wide. "Baby boy, 10 pounds, 9 ounces, 10 toes, 10 fingers, 1 banana and 2 walnuts." I guffaw at his references. "2 eyes, one nose?" He smiles, "Yep." I'm grinning, "You are something else Cook. Are we doing anything as a group for Caleb and Amelia?" Cook shakes his head, "No, we just let folks do what they want. Gets too complicated trying to track everyone down for money."

I look around, "Where's Nova?" Cook grins, "She's the doula." *Is there anything that woman can't do?* He looks at me and replies, "NOPE," and pops the P. *He acts like he is the proud papa.* Cook looks at me all stern, "How can you NOT be excited about new life?!" I reply, "I am Cook, truly. I'm just a wee bit jealous. I feel like I got gipped out of my childbearing years. I'm happy for them though!" Cook nods and pretends to understand. *Men just don't get it.* "Get what?" I hear and turn to see Malcolm walking through the door to the dining room. Cook asks, "Did you guys want supper tonight?" Malcolm motions to me, "I promised her I would cook." Cook grins as he is walking away, "Do you need anything from the kitchen before I close up shop?" Malcolm jogs after him, "Do you happen to have any fresh basil? The store was out." Cook hooks him up with a zippy bag of basil leaves and sends us away.

***

CHAPTER 16:

Malcolm grabs my hand and walks out the door to the warehouse. He takes me past the staging area and turns to the right to walk down the dark, wet hallway to the apartments. *Atleast I'm not alone this time.* It's pitch dark inside the hallway. I can smell and hear Malcolm next to me, but I sure can't see much.

All of a sudden, he grabs me and pushes me up against the wall. Before I can squeal, his lips are crashing down on mine. I open my mouth for a tongue wrestling match. And then it's over as fast as it started. My pulse is racing and I know I'm grinning stupidly. We walk a little farther and he does it again! But the kisses are over almost as fast as they start. *Ugh, can't a girl just have a minute's worth?* I can see the light from the glass double doors just starting to shine in front of us.

Malcolm lifts me up and carries me like a baby into my apartment. His lips are down on mine in a heartbeat as soon as I hear the door click shut. He carries me over to the couch and lays me out, barely breaking the kiss to lay me down. His body is fully pressed on top of mine and I can feel his erection straining against my leg. *Oh my, how big is he?* He lifts his head up to smile and his eyes are smoldering. I'm all up for a kissing match, so I grab his head and pull him back down to me. He is a gentleman and doesn't push me any farther. After several minutes, we are both so out of breath, we are panting. We sit up and smile at each other. Malcolm jumps up, "I'm gonna start supper." I can see his erection through his jeans. *Whoa.*

I stay put for a few minutes, feeling my lips to make sure they didn't fall off. Malcolm looks up at me and chuckles. "I'm going to make some noise babe." *Babe?* He's still grinning goofily, "I'm making chicken parmesan and have to pound these chicken breasts flat." "Yes dear," I call back. *I'm sooo not used to this... this domesticity. "Really? I've been waiting for this my whole life."* Malcolm sends back. He is whistling some tune I don't know.

I decide now is a good time to find the cat because I haven't seen him and normally, he runs out to greet Malcolm. I go around the corner into the bedroom and I hear Mr. Mittens, "Rawr, MEOW, MEOW, MEOW, Rawr." I call for him and still just hear him growling back. There's also a faint thudding sound. I walk into the bathroom and find him trying desperately to get out of the tub. But the water is running and he can't get a grip to pull himself out. He slips back down. I run over to him, "Buddy, how did you end up like this?" "Rawr, MEOW MEOW!!!" is all I get in return. *He must have turned on the taps in his panic.* "You poor thing!" I reach and grab a towel and throw it over him and wrap him up burrito style.

Malcolm comes into the bathroom. He holds his arms out, "What happened to my boy?!" Malcolm is glaring at me. The cat is glaring at me. *How did I become the enemy here?* Malcolm is holding Mr. Mittens, wrapped in the towel in his arms, like a baby.

He walks over to me, "Look honey, he's a purr-ito!" And Mr. Mittens is indeed purring like he has this man wrapped around his paw and he is proud of it. I roll my eyes so hard they might pop out of the sockets. *These two....* Malcolm is grinning. Mr. Mittens is purring.

I look at Malcolm in a panic, "What about the chicken?!" We both yell, "SHIT!" And run to the kitchen. It's fine, a little darker than we would like, but not bad. Perfectly edible. Mr. Mittens is indignant at the abrupt bouncy ride to the kitchen in Malcolm's arms. He struggles to get free. Malcolm lets him go and Mr. Mittens finds his spot on the couch for butt-cleaning time. I'm staring at him and Mr. Mittens looks at me, "*What Mom, it's already wet.*" *OH NO. OH NO NO NO NO NO NO. I'm not Dr. Doolittle.* I hold up both my hands and back towards the kitchen.

Malcolm is pouring the dry pasta into the pot to boil. He is laughing so hard, tears are rolling down his cheeks. I whirl around to him. "*You knew about this?! How does he even know English?! What in the...?!*" I'm bent over and holding my arms out to him in a pleading gesture. *Will all this crazy ever stop? Or atleast slow down?!* Malcolm sets the temp on the stove so the pasta doesn't boil over. He pulls a bar stool around to the kitchen side, grabs ahold of my shoulders and sits me down on it. He bends over a bit to look into my eyes. He picks up a piece of fruit and stuffs it in my mouth. "Now hush for a moment so I can explain." Mr. Mittens and I have been communicating for a couple weeks now. My eyes are wide. I'm still chewing and can't answer. *Smart man.*

Malcolm grins and then points back at Mr. Mittens. "That cat loves you more than air." OH, now I feel very vulnerable. I had no idea. *I thought I was just a food and poop servant.* "Mr Mittens and I have communicated about things that are best for you. He knows you better than anyone. I feel like his mind came around sooner because he wasn't injured like you were. Even though you received a rush treatment of the mineral, it still took time for your brain to heal and be able to receive messages."

Malcolm stops and takes a breath, "This is the part I don't want to tell you, but you need to know." He checks my eyes, "All of us have been able to hear your thoughts from the beginning." He checks on me again; I'm staring off into space. "Everyone has been able to hear them. I know what you thought about me the first time I met you. It was something about me being the most handsome man you've ever seen… or something like that." *Oh, wipe that smug look of your face, husband.*

I've come to. I have both my hands clasped over my mouth and my eyes are leaking. *It's only been a few days; how is this attachment happening so fast?* I'm in full panic mode at this realization. *This is not in my plan. Nope, not gonna happen. Nope.* I stand up, so fast that it knocks over the stool. I walk into the living room and start pacing. *No. No. Nope. Nada. Nope. No way.* I continue to pace, wiping at my face, wiping the tears away, pulling at my hair. I am absolutely panicking.

Malcolm carefully and calmly drains the pasta, puts it into a casserole dish, covers it, and sets it in the oven to keep warm. I don't know if he is freaking out because I'm definitely freaking out, but he shows me nothing but calm. He walks over to me like it's noon on Tuesday, and we are just having a chat. Mr. Mittens must have heard him because he takes off into the bedroom. Malcolm stands in front of me. I'm still rocking side to side. I've got my arms crossed over my chest, and the tears are rolling down my face. My expression must be of pure shock.

Malcolm takes the corner of his t-shirt and wipes my eyes. I'm still rocking. He wraps his arms around me and rocks with me. He doesn't try to control it; he just rocks along with me. Finally, I crumple into his arms and sob. I cry so hard it feels like it's coming up out of my feet. He sits down cross-legged on the floor. He arranges me on his lap. He's still rocking me gently from side to side. I can't even think. He is patting my hair, "Shhhhh darling, you're ok. Shhhhh…." I'm wrecked. I have no idea how long this goes on, but I'm past feeling anything. I've stopped sobbing. He carries me to the bed and tucks me in.

I found this out later: There's a knock at the front door. It's Cook and Kathleen. Kathleen has a syringe in her hand and asks, "Do you think she really needs this, Malcolm?" I hear him reply, "No, not now. She finally calmed down." Cook grabs Malcolm's shoulder, "Are you ok, man? She's had a lot going on in a short period of time, but so have you." Malcolm shakes his head no, looks at Kathleen, and points to the bedroom, "She's settled in there." Malcolm and Cook continue to talk in the front doorway.

Kathleen knocks on the bedroom door very gently. "Ellie, you awake?" I roll over, trying to free my legs from the blankets, but fail and end up half-wrapped up in them. She steps forward to help me sort myself out. Kathleen sits down on the edge of the bed. "Another big day. You've had a lot of those lately." I nod. My eyes are leaking again. *How is there any moisture left in there?* She giggles a little and asks me, "Do you need anything? Do you want anyone around?"

I shake my head no but hold my hand up. I grab ahold of her hand and whisper, "I don't want Malcolm to leave." She smiles, "I don't think he would leave if you tried to make him." I smile slightly. She stands, walks into the bathroom, and comes back with some face cream. She dips her finger into the pot and holds it up. "If you don't put some of this on, your face is going to be red and dry across your cheeks in the morning." I nod, and she wipes it onto my face, leaving it a bit thick.

I look up at her with weepy eyes. "I have a headache." She smiles again, "Ellie, we could all hear you crying. As hard as that was for you to let out, it was equally as hard for us to hear and see the images in your mind." *Oh no.*

She holds her hand up, "Stop. Why do you think we are so close-knit, like a family?" I shrug. *I've never really had a family.* Kathleen puts her finger under my chin, "We know what is going on with each other. It's unavoidable. We all develop skills at keeping things to ourselves, but for the most part, our private lives become part of the dynamic here. I can tell you that I felt the pure despair of what you were feeling from my house. It was that LOUD in my mind. I can only imagine what Malcolm felt while holding you. I'm sure it was very close to what you felt." I groan. *I don't want to cause anyone pain.*

"It's not pain, Ellie. It's understanding that we share with each other." She sighs, "That's the beautiful part of it. I firmly believe that humans, as a species, were meant by God to share thoughts and feelings intuitively. We are blessed to experience that here. It's a rare gift." I slump up against the headboard. "I need to ponder this some more, Kathleen. I can't take in anything else tonight. Malcolm told me earlier that he's been talking to the cat. And Mr. Mittens talked to me. Now, all this, and I gesture out the door."

Kathleen giggles, "That Mr. Mittens is a sly one. He would always try to sneak out the door of the infirmary to get to the dining room with me. I could hear him and would catch him and put him back in the bed with you. But I always told him I would bring him back a treat, and I held up my end of the bargain. Mr. Mittens loves you deeply." I nod. "That's what I hear." She gives me a sip of water and tucks me back in. "Thank you, Kathleen," I say sleepily.

###

I wake up the next morning with strong arms around me. I'm lying on his chest. *There isn't an ounce of fat on this man; how am I going to be able to snuggle comfortably?* I shift, and as I do that, he wakes up. "Huh?" I smile up at him, "Hey baby, good morning."

He grins, "Hey, baby." And falls back to sleep. I decide that after the night I had and basically keeping everyone awake and in pain with me, I should probably take a bath. I start the tap running and see Mr. Mittens run out of the bathroom like he's lit on fire. *Looks like he learned a lesson.* I sink into the hot water and find that my cheeks are tender to the touch. I'm glad Kathleen thought to put the face cream on for me. *How will I ever repay that woman?*

Malcolm walks in. "You want company in that bath?" I hold my hand up, "Only if you can remain unfrisky." He nods. "I'm not feeling frisky in the slightest." He strips off right in front of me. *Oh, how I wish I felt better. Because any other day, I would pounce on that.* I sit forward to let him sink behind me in the water. He whispers in my ear, "Pounce, huh?" Something else decides it wants to be frisky, and he looks down, "Down, boy." He pulls my body up against his and reaches over to push the button for the jets to run. I lean up against him, my back to his chest for a long time.

The water stayed warm. *The pump must have a heater in it.* Pretty soon, the jets turn off, and the bathroom becomes quiet except for the occasional swish of water. I look up at him, well, his chin anyway, and ask, "Have you heard from your Dad and brother?" He kisses my forehead and replies, "Not yet today. I forgot my phone in my office. We will need to get to it soon. Until then, one of the guys can answer it for me." I stay still and quiet for a while.

Finally, I say, "Can we go get ice cream?" Malcolm laughs and agrees. He lets me get out of the tub first. I pretend I don't see him eyeing me. He stands up out of the tub with his penis jutting out in front of him. *Drool... this hunky man wants me?* He grins and wraps a towel around his waist. Then he grabs another one off the towel warmer and wraps it around me, rubbing the towel all over my arms and grabbing my butt. I tilt my head to look at him. He lets go and holds both hands up.

The towel drops, of course, and he looks at me up and down again. I swat at his rear, "You're terrible." "No, babe, just horny and in love." He replies with a grin. *Now he's in love... These crazy days....* We get dressed and I wear the same thing as the day before because that's the only clean clothes I've got. We stop at his apartment so he can change.

###

Malcolm walks me out of the apartment hallway, down the dark corridor, and through the warehouse to Chef's dining room. I look up at him. "No, Cook is the pie expert, Chef is the custard expert." We walk in, and Chef strides towards me fast. He wraps me up in a huge hug, then holds me at arms-length to look at my face. "You DO need custard. I don't even care that it's 10 am."

He walks around behind the front counter to a row of refrigerators with freezers on the bottom. Malcolm sits me down on a bar stool and hands me a napkin. Chef comes back with a generic plastic bucket, sets it down hard on the countertop, peels back the lid, and hands me a spoon. I roll my eyes. "I'm not going to eat the whole thing. Could I please have a bowl?"

Chef comes back with a red bowl and scoops 5 huge scoops of custard into the bowl. He asks, "I'm guessing you're a plain French Vanilla girl?" I hang my head, "Yes, sir." Chef laughs, "Oh no, I totally get it. There's a lot of nuance in just vanilla. Sometimes the toppings overpower that, and the effect of the silkiness of the custard is lost."

Malcolm looks stunned, "You mean to tell me I've been eating your custard wrong all this time?" Chef looks directly at him, "Yes, you have been treating it like ice cream. I've been insulted but haven't said anything." Malcolm's mouth is wide open in an O. I take a big spoonful of the frozen custard and stick it into his mouth. I'm giggling. Chef dips up a bowl for Malcolm and another for himself. Looks like it's frozen custard o'clock at the quarry. Pretty soon, the warehouse employees have trickled in, and then the office staff and even the harvesting crew stop by. By the time everyone has dipped a bowl, we have decimated Chef's supply. There might be a spoonful left for Mr. Mittens.

There's a crackle from a radio over by the stove. "Tater and Bulldog to Base, do you copy?" Chef laughs and turns around to answer it. I can tell it's Nathan's voice. "Base to T and B, we copy. What do you need?" The radio crackles again, "Tater and Bulldog just wanted you to know it's not fair, and we want our share of custard when we get back. Over and out." Everybody laughs. I turn my head to Malcolm. "They could hear us?" Malcolm just smiles and takes another half-melted bite of custard.

When my bowl is finally empty, Malcolm takes it from me and hands it to Grant, who is working feverishly to get all the bowls and spoons run through the dishwashers before lunch. Because, of course, today's lunch is some fancy type of stew, and Chef needs all these bowls to serve it.

Malcolm directs me out of the dining room. I wave to everyone as I go. Almost all of them nod, smile, or wave in return. Only a couple sets of haunted eyes watch me. *Oh no, I've scarred them for life.* Malcolm replies out loud, "No, you haven't. They are just worried that you experienced more than what you were able to release last night. They are hoping you don't have to go through all that again to finish healing." "Oh." Is all I can say.

***

CHAPTER 17:

Malcolm settles me at my desk, "Let me check on things here, and then we can go for a ride." I nod; he kisses my forehead and walks away. I turn to my computer and start working on all the new employees. *I wonder how that is going to work. Are they able to get a warehouse set up that fast?*

Maia is missing again today. Terrell walks over to my desk. "She rode with Bryan to the reservation to check on things there." I nod and start to turn back to my desk.

He stops me, "Ellie, can I give you a hug? I had no idea." I nod again and stand. He walks around to me and wraps me up in a big bear hug. I'm so short compared to him; I'm basically hugging his stomach. *We gotta shake off this atmosphere around here today.*

Right on cue, Monkey walks into the office and tosses me the keys to Malcolm's big truck. *Oh really.* I grin mischievously. Terrell chuckles, "I'll leave you to it. You better not scratch it; he will have a meltdown." I run for the front door behind Monkey and hop into the truck as fast as I can. Terrell is standing in the front window, watching me. I start up the truck; it's a stick. *Drat.* I can drive a stick, but I really don't want to hurt the truck since I'm not used to whichever transmission that's in it. I just pump my foot on the fuel pedal, keeping my left foot on the clutch so I don't kill it. The truck growls loudly, and I can see Terrell laughing through the window.

Here comes Malcolm at a trot. He opens the driver's door; “Over, woman!” motioning to me to scoot. I don’t go all the way over; I stay right in the middle to harass him. He has to reach between my legs to shift now. *Oh, all the metaphors...*

Malcolm backs the truck up, turns around, and takes me back to the storage container in the woods near his new house. He helps me down out of the passenger side. “I thought you might like to have more of your clothes.” I sigh in relief. I haven’t felt like myself without all my clothes. *Whoever packed all this up took their time.* Everything is neatly folded and packed into totes. *I hadn’t realized how much of a clothes whore I am.* Malcolm helps me load totes into his truck until the entire bed is stuffed full, 2 layers deep. That isn’t all of the totes marked as “clothes,” either. Malcolm grins at me as he shuts the tailgate, “I think you’re going to need both the closets in the apartment just for yourself!” *“You THINK you’re funny.”* I send to him.

He picks me up, slings me over his shoulder, and deposits me back In the passenger seat. I look at him with doe eyes. He’s suspicious, “What do you want?” I raise my head and ask to see the house again. He grins wide like I made his day.

###

Malcolm drives up the path to the house - not much has changed. I hop out of the truck before he can get to me this time. I walk all the way around the house and find that he has built the house into the side of a hill that leads down into a dip in the pasture. He points, “There’s a spring at the bottom down there. We will have all the fresh water we want." *Oh really, WE is it now?* He looks up at me kinda sheepish but doesn't say anything.

The basement walks out into the area surrounded by cool trees. There is a deck above, so you can walk out under the trees from the first floor as well. "Oh, this is going to be lovely in the summertime. Is there a flat spot for a garden?" He points east and I see a large flat area about 50 feet from the house. "I think there's good dirt under that grass over there." I chime in, "Do you have any cow poop?" He starts laughing, "Um… why?" I look at him with a stern expression. "You have to have good cow poop to grow a good garden. Pop Pop always had a great garden, and he would spread a thick layer of poo from the cow lot on his garden every fall." He looks at me and tilts his head. "We will take Pop Pop's advice then."

Malcolm digs his phone out of his pocket and makes a call. "Yeah, Cody? We need a good trailer load of cow poop. What would be the ETA on that? What do you mean it's heavy? Well, call Bryan and the shipping company and find a trailer he can haul it with." He looks flabbergasted. "I know it will take a lot of fuel to haul it. Yes, I need it at the house. Cody, don't give me crap about this." I can hear Cody roaring with laughter on the other end of the line.

Malcolm still looks perplexed. "Yes, Bryan can make a second trip out. It's not like we can't rent another trailer!" Malcolm hangs up on Cody. I'm grinning because I know what that man is thinking. I ask as gingerly as I can. "Malcolm, exactly where is the cow poop currently located?" He looks at me like I'm dumb. "Wyoming, where else would it be? I don't have any livestock here yet." I shake my head and start back for the truck. *This man is just too much.*

We arrive at the office, and Monkey meets us with a couple guys and ATVs to match. As soon as the truck is parked, they open the tailgate and start unloading the totes of clothes into the ATVs. Within an hour, all the totes are inside the apartment, and the guys are gone. Nova arrives with boxes of hangers. *Why didn't I think of that?*

She smiles at me, "Because you've had a lot going on." It takes us 2 hours, but we are able to get everything organized and re-hung. Cook shows up with 5 pizzas for supper, like he is going to feed an army. Malcolm walks in right behind him, and I understand the need for that many pizzas once the guys polish off a pizza each. Nova and I finish one and a half pizzas by ourselves. Cook wraps the remaining pizza up and sticks it in the fridge. "You'll need the extra for snacks." He lays his arm around Nova's shoulders and leads her out the door.

Malcolm and I are alone in the apartment again. *I do NOT want a repeat of last night.* He nods and takes my hand. He leads me to the bed and lays down, patting his chest for me to lay on it. "Just come snuggle with me." He says, and we fall asleep almost instantly.

It's the middle of the night, and I wake up, startled to find Malcolm in my bed again. We are fully clothed, so there is nothing to worry about. I don't remember seeing Mr. Mittens when we got back. And we didn't see him all evening either. *Where is my cat?* I rouse Malcolm. "Wha… What honey?" He kisses my forehead and opens his eyes enough to see my worried face. I blurt out, "I didn't see Mr. Mittens last night, did you?" He focuses his gaze on the wall like he's thinking hard. "He isn't here, Ellie. We better go find him." "Surely, he just went out into the hallway," I say, but my heart is sinking. *The guys had the glass doors open for a long time, hauling in the totes.* Malcolm sits up and startles me. "You're right. Let's go."

I slip my sneakers back on, and Malcolm runs down to his apartment to grab a radio and his hat. Now he's wearing a blue plaid flannel shirt that's open down the front, wrinkled jeans, black scuffed-up cowboy boots, and a black cowboy hat. If I wasn't so worried about Mr. Mittens, I would turn into a puddle right here and right now. *How'd I get so lucky?* The guys didn't leave an ATV by the glass doors so Malcolm leads me through the dark hallway using his phone for a flashlight.

I ask, "Can you hear him at all?" Malcolm shakes his head no, and we keep walking.

###

Once we arrive at the staging area, he loads me up and we take off down the corridor that leads to where I got stuck. Once we are in range, the radio crackles. "Tater and Bulldog to Malky, over." Malcolm chuckles, "I haven't heard that in a very long time."

He speaks into the radio, "Malky and Trouble to Tater and Bulldog, we are on the way. Searching for a lost nosey cat. Over."

There was no reply, and I'm getting very nervous about Mr. Mittens' well-being. *He's going to think I don't love him.* I swat Malcolm on the shoulder, "Trouble… huh?" He replies, "Yep, you've been nothing but trouble since you first showed up." And he gives me the widest grin he's got. *Harrumph anyhow.* We are speeding down the corridor until we get into some muck that is several inches deep.

The mud is splattered all over the ATV and us. Pretty soon, the corridor exits into another room, but it's structured more like a normal cave. *There's stalactites!* I marvel at the tall cathedral ceiling, all covered in dripping icicle-like structures. We've descended into the water now and are following an underground creek through the cavern. We turn a corner and see Nathan and Mr. Louis standing together along the side of the creek. Nathan is holding a pissed-off, very wet, and muddy cat by the scruff. Mr. Mittens has murder in his eyes and keeps swiping at Nathan as best as he can from his precarious position in the air. *He's never liked Nathan. I don't like Nathan very much right now, either.* Malcolm's eyes soften at the sight of Mr. Mittens.

"Buddy, what were you thinking? You're a fright!" Nathan drops the cat into Malcolm's arms. Mr. Mittens snuggles right in. Mr. Louis is laughing out loud at the scene, looking from one brother to the other. "He has you boys pegged to a T!"

I walk up to Mr. Louis and hold my arms out for a hug. "Thank you for finding him, Mr. Louis. Have you had very much luck finding more sapphires?" Mr. Louis pulls out a tiny leather bag from his top pocket. He looks over at Nathan, "We found a few more. Hold your hand out."

He pours a small fortune in gemstones into my outstretched hands. Nathan walks over, pulls an identical bag from his top pocket, and pours his stones out into my hands as well. My eyes are wide. Malcolm walks up and slaps Nathan's back with his free hand. Mr. Mittens growls.

Malcolm smiles down at Mr. Mittens, "Buddy, you're one lucky cat. I wouldn't growl at him anymore." Mr. Mittens looks at Malcolm with wide eyes. *He just understood every word.* I remind everyone that I'm still holding all these gems, and Nathan scoops a few back into his bag and motions for me to put the rest in Mr. Louis' bag.

I look up at Mr. Louis, "Are you guys planning to come back anytime soon, or are you going to start living in dark caves from now on?" Both of them laugh. Nathan replies, "We are getting low on supplies and we're tired of eating our own cooking on a camp stove. We will probably be back tomorrow evening. There's one more pocket to investigate." I narrow my eyes at him but behave. "You both be careful, and we will see you tomorrow." I pat Mr. Louis on the shoulder, and the 3 of us load back into the ATV. Malcolm has me drive since he can't let go of the "baby." *Geez.*

###

I drive the ATV right up to the glass doors. Malcolm carries Mr. Mittens straight into the bathroom. I follow and strip off. Malcolm gapes at me. "What? I gotta get him cleaned up. I might as well clean up, too." He agrees and hands me the cat.

He strips off and walks over to the shower to get hot water coming out everywhere. *MMMM HMMM, walk right in front of me with that gorgeous naked ass.* He walks to one end of the shower, pulls out a solid-glass pocket door and motions for me to come inside with Mr. Mittens. He slides another door shut behind me. *Well, isn't that handy?* Mr. Mittens is pissed. He walks through the shower from one end to the other, struggling to get out. With the doors shut, he can't, and Malcolm and I can clean up without worrying that Mr. Mittens will escape before he is clean.

I get done washing all the important bits and realize my hair is caked with mud. I grab the bottle of shampoo, and Malcolm says, "Here," holding his hand out. I can barely hear him over the rush of all the water. *"Here."* I hear in my mind. *"See how useful this is?"* I nod. *I haven't gotten the hang of everything the intuition can do.* Malcolm works the shampoo through my hair and takes the hose wand off the wall to rinse it out. Then he holds me from behind for a long minute. I can feel his erection against me. *"I'm not sure I'm ready for that yet. I love how comfortable we are together, though."* I send to him.

"There is no rush, baby. We have the rest of our lives. We can take our time. It will be sweeter when it's not rushed anyway." He soaps up my back, and I wash his back and slap his butt. We both reach down to soap up Mr. Mittens and rinse him back off with the

hose wand. I send, "*Somebody is gonna have to pick up the pissed-off soaked cat.*"

I look up at Malcolm with wide eyes. He smiles, and Mr. Mittens practically jumps into his arms. *It's a bromance.* We all 3 step out, and I grab a warm towel to wrap around Mr. Mittens. Malcolm gets the cat swaddled while I hurry up and dry off. I hand him a warm towel so he can do the same. We have made quite a puddle, so I lay my towel down on the floor and reach for another one. All 3 of us head back to the bed. Neither one of us put on a stitch of clothing. I want that man so bad I can taste it. *But I'm just not ready.*

I don't know what time it is, but I know it's morning, and not a single one of us moved during the night. I look over at Malcolm. He is on his back, the cat curled up in the crook of his arm. They are both still sawing logs. I stop in the bathroom to pee and put my robe on. It feels so much better to have my clothes. I pad out to the kitchen barefooted. The pizza is still in the fridge, but I would like a more substantial breakfast than leftovers. I don't feel like going to the dining room this morning. *This whole domestic thing is growing on me.* I dig out the eggs and bacon. There are some potatoes in the cabinet I can chop up. I look over towards the door and see a slip of paper under it. I walk over and grab it, a note from Nathan, "Wanted you both to know we decided to quit while we were ahead and came back. See you in the office. N" Well, that's a relief. *I hated leaving them there last night. Well, Mr. Louis, anyway.* I chuckle at myself. I start frying bacon, fire up the coffee maker, and chop a few potatoes. *Why do you only eat 1 baked potato with a meal, but if you chop potatoes up, you have to serve 2 per person to be satisfied?*

Malcolm walks into the kitchen, naked as a jaybird, carrying his cat in his arms. His erection is jutting out right in front of him. *How did we get to this already?* His brown hair is all a mess. He scratches his head with his free hand and yawns loudly. "I smell bacon. Do you smell bacon?"

I smile at him. "Yes, dear, I made you bacon." I set a plate down on the bar top. I motion with my hand at his nakedness, "You gonna eat like that?" He grins, "Why not? Turns the kitchen girl on…" I blush. *Yeah, it does.* I sit next to him with my own plate, and that's when the coffee maker decides to finish gurgling. He puts his hand over mine, "I've got it." He stands up, grabs 2 mugs, and fills them to the brim with hot black coffee. *My coffee is delivered to me by a naked man, whom I got involved with just a few days ago, in an apartment that isn't mine, that I'm being forced to stay in because my house is rented to people that I don't know.*

Malcolm sets his coffee mug down and holds my face in between his hands. "We are going to get everything sorted. Nathan knows he made some major mistakes and overstepped his bounds. He told me the gist of your story. I got the rest of the story from you the other night. I'm not going to push you. As for the house, you can live in mine when it's finished if you want." I look up at him, eyes wide.

He continues, "I know Robert and Darlene have a year lease on your house, and it would be up to you to break it as the homeowner. They are good people, and the less things get nasty with them, the better." I sigh heavily. "You're right. Do you think Nathan will be able to sit and talk to me about all this?" Malcolm laughs and almost shoots coffee out his nose. "Do you think you can be civil to him?" *That's a damn good question.*

***

CHAPTER 18:

We get ready for the day in our respective apartments. He walks down the hallway in just a towel – *oh, how I love watching him walk around in that towel.* Just knowing I could rip it off him and have the whole meal deal - *it'd be so satisfying*. I go through my clothes and select a pair of navy slacks. *Nope, I think I'm going to wear a skirt today.* The weather is still nice; it is heading towards fall, but the days are warm. I find a flowy blue floral skirt from years ago that I had completely forgotten about. I add on a cream-colored twin set and tan lace-up sandals. I look in the mirror. *That works.* Malcolm walks into the bathroom just in time to see me making my mascara face. He is dressed in a pair of navy pinstripe pants and a light blue shirt. His shoes are brown and not scuffed. He is carrying his suit jacket over his arm. He looks me up and down and tilts his head. He sends to me, "*Are we going to church?*" I answer out loud after I finish my mascara. "It's a Thursday." *I think.* Malcolm grins, "Wrong. It's Tuesday. We've all been working overtime to get Oklahoma set up. That's why you haven't had any weekends." *I haven't had any periods either. I'll need to talk to Kathleen about that.*

Malcolm looks into my eyes. "We can talk about that. Your body is going through a major renovation on the inside. You might miss several periods. We both know it isn't because of sex. I need to tell you Mom's story sometime." He holds out a hand, "Let's get going." Malcolm kisses his cat, and we step out the door. Terrell steps out of his apartment at the same time. He looks glum. Malcolm makes small talk with him as we all walk to work together.

Terrell peels away and goes straight to his office, shutting the door. I drop Malcolm off at his little office and walk to my desk. With Maia still gone on the run to Oklahoma, I'm answering phones, again. The morning slips by. *Man, it feels good to sit and get some of this work done.*

I look up when I see Malcolm standing over me. "Hey, Sweet Cheeks, it's lunchtime." He holds out his hand for me to take it. I look at the two of us, all color-coordinated, and we really do look like we are going to church. He leads me to Cook's dining room. Today's lunch is beef and noodles. I opt for a side salad to go along with it. Nova and Cook are swamped up at the front counter, so Malcolm grabs our plates, and I get the drinks. I look over at him, "We HAVE to get them some help. The more people we add, the more they are feeding every day. And Nova cannot continue to do 2 jobs." Malcolm sighs, "You're right. Let's stop in this evening and talk with them about what they want."

That's one thing I do love about the structure here: everyone wants to work; we are a close-knit family. It's so refreshing to have a choice in how things are done and how we want our work to expand and contract. We also have a choice in what type of work each of us do and are free to move between areas if we want. If I wanted to start cooking instead of scheduling, I could just make that choice and find a place in the dining room where I fit and go to work. It does create some workload nightmares from time to time. But knowing you don't have to remain stuck in a role you're bored with and knowing you won't be demoted for choosing another path really helps maintain your self-worth.

###

When we return from lunch, Terrell's door is still shut. I send to Malcolm, *"He feels very closed off to me."* Malcolm nods and says out loud, "He has shut everyone out." My mouth forms a little o. "You can do that? Shut down your mind so other people can't hear it?" Malcolm replies, "Yes, it's kind of like shutting the blinds and deciding how far to turn the wand to let more light in or out. Right now, he has room darkening blinds up and they are shut tight."

I sigh. "Terrell is really torn up about all this, isn't he?" He smiles and looks down at me while I sit. "He had big plans for the 2 of them. Everything has changed now, and he doesn't know where he fits. Nobody has answers for him, either. It all has to play out on its own." He leans down and kisses my forehead. "I'm going to be on the phone all afternoon, and then I have to meet with Dad and Nathan. Would you mind if I meet you in the dining room later?" I nod and turn back to my screen, "That's fine, babe." As I said that, a mega-watt smile formed on his face. *I think I just made that man's day. "You sure did."* I hear in my mind.

###

I'm not paying attention and work way past 5 pm. The only thing that alerts me to a break is my bladder. I race to the restrooms and see Malcolm, Nathan, and Mr. Louis all seated in the dining room. I hear "Ellie!" from somebody, but I have to pee and pee now. Once I solve that problem, I head to the dining room and find an empty seat between Nathan and Malcolm. That leaves me facing Mr. Louis at the small round table.

I smile at him. "Mr. Louis, how's the sapphire business?" He grins and pulls a small velvet box out of his top pocket. He opens the box for me and inside is a fairly large, rectangular cut sapphire that is absolutely gleaming in the low light of the dining room. I gasp, "That's amazing!"

Mr. Louis looks over at Nathan and smiles widely at him. "I think we both needed that time in the cave." Nathan sighs, "Yes, Dad, we sure did. I really enjoyed hearing so much more about your childhood and early farming days. It's a treat to physically work beside you." Mr. Louis' eyes begin leaking a little. I chime in, "You had time to bond again?" Both of them nod. *Nathan does seem calmer somehow.*

Malcolm brings up the Oklahoma location, and they start chatting about logistics and building plans. I gather they are able to rent a warehouse with office space for the time being. I overhear Malcolm saying, "Terrell is tracking down a food truck for them so they have a kitchen until we can get everything up and running…."

I turn just in time to see Chef walking in. He is carrying a big tote full of food containers. I stand to help take the leaning container off the top. He looks at me in relief. "I've worn everyone out today. Grant gave up, and Cook and Nova are whooped. It's just me feeding you stragglers tonight."

I look up at him; he looks tired, too. “Chef, I can help you. What is all this? Tell me what to do.” He smiles, “These are the leftovers that Malcolm fixed the other night that you didn’t get to eat. I didn’t want it to go to waste, and I figured there was just enough for the 4 of you. It just needs to be reheated.” I put my hand on his shoulder. “I’ve sat at a desk all day. Would you mind if I took care of this? You need to go rest, too.”

Malcolm walks up to us. “Chef, let us handle this tonight.” Chef shakes his head but hangs it and walks away. Malcolm looks over at me, “We have to talk to Dad and Nathan about the food situation tonight.” I agree with him, and we set to work on reheating the chicken parmesan. I find some salad leftover from lunch in one of the refrigerators, and we exit the kitchen with 4 plates piled high. Nathan stands to help get the plates set on the table. Malcolm immediately starts to talk, “Dad, we need more kitchen help and probably more dining space, too.” Mr. Louis nods and dives into his food. Nathan agrees with Malcolm and sets right to eating. *Well, that was easy.*

I decide to break the silence. *They were all eating so quietly, too.* “Has anyone heard from Caleb? Is Amelia doing well?” Nathan looks at me, “Ellie, I think we all forget that your Intuition skills haven’t fully blossomed yet. Caleb let us know that Amelia and the baby are doing great. He plans to start appearing back in the office tomorrow morning.” I realize what he means and reply, “Oh, he sent a thought to everyone and I haven’t been able to receive it.”

Mr. Louis points at me with his fork, "Precisely, but you will get "tuned in" soon enough."

I grin and go back to my food. Malcolm sits back and places his hand over his stomach. "I'm a pretty decent cook, ya know it?" All of us chuckle and he waits for us to finish our food. I stand up and take the plates. I'm hoping there's some pie left in one of the refrigerators; I look through 3 of them and then. *Score!* I plate up 4 slices on paper plates and hand them to Malcolm to take back to the table. I don't want to leave a mess for Cook, so I place all the dirty dishes in one of the industrial dishwashers before I leave the kitchen. We polish off the pie, and Mr. Louis and Nathan head back to their offices. Malcolm leads me by the hand to the good hallway.

"I'm going to stay and talk with them about what we need to do for a better meal set up. I will see you in the morning?" I nod and kiss him lightly on the mouth. "See you tomorrow, babe." I can feel his grin behind me as I turn around to walk down the hallway.

###

After a few steps, I turn the corner and intercept Terrell again. He is on his cell phone, walking slow before he walks into the cliff face portion of the hallway. "Yeah, baby, see you in a couple of days. I'm missing you something terrible. Love you." He hangs up and grins at me. He still looks so tired.

"Hey Ellie, that was Maia. She is still riding with Bryan. He has to run up to Wyoming for something for Malcolm, I guess. It will be a couple more days until she is home again." I grin; I think

I know what that "something for Malcolm" is. *Won't Bryan and Maia both be surprised at their smelly cargo...* Terrell wrinkles his nose, "Oh no way Ellie." His eyes widen, "Oh, you're not joking!" And he busts into laughter. "Of all the things to have hauled that far..." I smile up at him and we walk together, "I know, I shouldn't have said anything to him about it; I didn't even know there were cattle in Wyoming."

Terrell slings his arm around my shoulder, "You're going to find out those 3 have their hands in a lot of different industries. They will never run out of money with all the different ways it pours in." He bids me goodnight at his apartment door, letter M. I point to it, "Oh, for Maia." He looks at it, "Oh geez, Ellie, I never thought of that. What a great analogy." I smile and wave as I walk away.

###

Mr. Mittens hears me approach the door and is already meowing at it. I grab him as soon as he tries to sneak out. "Oh no, you're not." He looks behind me for Malcolm. "You're not going to get to see your lover man tonight either." I set Mr. Mittens down and he skulks back to the bedroom like I have totally ruined his night. I roll my eyes. *What a deal.* I decide it's a good night to take a long hot bath and even pull up a book on my phone to read.

I hear footsteps in the living room. Mr. Mittens leaps off the bed and "Meow-ow-ows" his way to meet Malcolm. I can tell it's Malcolm just from the cat's meow. *Pitiful.* Malcolm stands in the doorway of the bathroom, holding his cat. "I could tell you were still up." He walks over to me and kisses me deeply on the lips, and a sigh escapes my mouth.

"Yeah, but I'm heading to bed here in a minute." He leans down again. "I just wanted a goodnight kiss. I'm going back to my apartment tonight. I don't want this to become a habit until we are both ready for it to be." I'm disappointed. *But it's probably best.* He kisses me deeply again and walks out. I wait to get out of the tub until I hear the front door click.

Mr. Mittens settles himself back into his nest on the bed. I take the time to braid my hair before bed so I don't end up with a wet rat's nest in the morning. I settle in and fall right to sleep.

###

We both wake up to the smell of coffee in the morning. Malcolm snuck in and started the coffee pot for me. There's a note, "Breakfast is in the oven. See you later today; we are meeting with the contractors for the new dining hall this morning." *That was fast.* "Miss you. Love, Malky." *Does he really want me to start calling him that?*

I reach up and turn off the oven. What I find inside is on par with the breakfast Cook prepares for me. I think the only thing missing is the ham. I slice off a bite of egg for Mr. Mittens. He looks up at me and licks his chops like he would have preferred more. "Sorry, Buddy, I'm finishing the rest myself." He turns around, sticks his tail up in the air, and trots off. I hear, "*Malky will share with me later."* If I had something soft to throw at him, I would. *Brat.*

***

CHAPTER 19:

I finish getting ready for the day and walk to the office. There's a trio of tall, muscular guys standing in the front of the office. One of them is smiling menacingly at Maia, like she would be a treat. Terrell isn't here yet. I send off, "*Terrell... get your butt up here!*" Maia is being polite and staving off the jerk's innuendos. I look all around the office and don't see any of the guys, not Caleb, Malcolm, Nathan, or Mr. Louis. *This could be a problem fast. Where is everybody?*

Just then, I hear quick footsteps coming up behind us. It's Terrell. *Whew.* He smiles and shakes hands with all 3 of the guys. He introduces himself and has them all step outside. He is standing out on the front porch with them, trying to make conversation. "*Malcolm babe, you better get up here too!*" I hear the door to Nathan's office open and bang against something. Malcolm trots up front and goes right out the door. "Great to see you guys again!" *Whoa, that was close.*

Maia turns to me with wide eyes. She whispers, "Could you hear what that guy was thinking about what he would do to me?" I shake my head no. "I couldn't, but I could imagine what he was thinking by how he was leering at you." She looks around and continues to whisper. "How did Terrell know to get here so fast?" I point at my head and say, "I told him we needed him." Her mouth drops open. "Ellie, that means you're developing more skills! I didn't hear you ask him; you sent it directly to him! Did you ask Malcolm, too?" I grin a little sheepish, "Yeah, I guess I did." *How does that work now?*

I keep my focus on Maia. "Explain to me how I did what I did. You're telling me I sent individual messages to particular people without anyone else overhearing them?" She nods and replies, "Yes, that's how we send messages to people back and forth around here. If Malcolm needs Terrell in his office, he will just send him a message that way." *Well, isn't that nifty?* She nods again, "It really is and I believe it helps this organization run smoother than most."

There's a whole bunch of footsteps on the porch, and we see all 5 guys walking across the parking lot to the front and right of the front door. I can't see Malcolm anymore. I turn back to Maia; I guess we are getting a new dining hall. Nobody seems to waste any time." She laughs, "As soon as a need is found for something, Mr. Louis acts on it. Having a lot of connections really helps things move faster."

I turn around but have a thought and spin my chair to turn back to Maia. "How long has this office been here?" Maia goes into a diatribe about how small the office area originally was over by Chef's kitchen, and that space was added to the dining room there. The warehouse is original as well. This office space is newer, about 3 years old. I smile at Maia, "I'm betting Malcolm's NEW office will end up being the OLD dining room; what do you think?" Maia laughs, "I think you're right."

The phone rings, and I can hear that loud man from Texas on the other end of the line. He is not happy, as usual. Maia interrupts him, "I don't care what your opinion is of adding an Oklahoma location, Jim. You are on the opposite side of Texas, so what difference does it make to you?" Maia is grimacing. She interrupts him again, "Jim, you need to take this up with Mr. Louis, not with me. I'm just the receptionist. I don't understand how this affects your operation in Texas at all." She hangs up. Maia points at the phone, "That's a problem."

I nod and go back to my computer. Malcolm walks back in with a set of construction plans under his arm. I see the remaining 4 guys walking across the lot. Terrell grabs the guy who was hitting on Maia earlier and socks him right in the mouth. The guy falls but tries to get back up and Terrell punches him again. *I bet that felt good.* Terrell walks in through the front door and shuts it hard, just short of a slam. Malcolm turns to look at him. "Do you want me to request he doesn't stay on the crew?" Terrell shakes out his hand. "No, I think I taught him what he needed to know." And he walks into his office and shuts the door again. Maia just watches him go. Malcolm's face is barely hiding a grin. He turns to me. "See you for lunch?" I nod and turn back to my computer again. *Wow.* Maia giggles, and I giggle to match. *Manly men.*

###

It's about 11, and something tells me I need to check on Cook and Nova. Oh yeah, they are already getting busy. I roll up my sleeves and jump in. There's a pile of dishes that need to be put in the dishwashers, so I start those; then I head to the trash cans and get them emptied. Malcolm walks in, "Ellie, I was…" His face falls, "Looking for you." I jog past him and smash a kiss on his cheek as I go by. I yell over my shoulder, "They need our help!" Malcolm rolls up his sleeves and joins Cook at the stove.

By the time we have everybody served, we are famished and worn out. Mr. Louis toddles in to find his workers all mixed up and hanging their heads. "What's this? Malcolm? What are you doing here? Ellie, you don't have any business working like this since your injury!" He takes a breath, "Cook, Nova, I will have this evening's dinner catered. You're taking the night off." Just then, Nathan walks in, "Whaddya need Pop?" "We have a problem, son." Mr. Louis replies.

Malcolm kisses me on the forehead and sends me back to my desk. Mr. Louis stops me mid-stride, "Ellie, call this catering company and tell them we need a service for 200 for 6 pm. Dishes and all, they can handle it." He hands me a slip of paper. "You go ahead and make all the food choices, I gather you will know what goes together."

I nod and walk out the door. Out of the corner of my eye, I see the 3 of them gather around the small table again. I sit down at my desk, still holding the paper. I look at Maia. She points, "What's that?" I fill her in on what just happened, and her eyes open wide. "I will go ahead and call the food trucks in for lunch tomorrow." I look up at her. "Has anybody informed Chef and Grant of this?"

She looks at me sideways, "Yes, they already know." She points to her head. *Duh.* The guys would have sent messages to Chef using their intuition. I get ahold of Cathy's Catering, and after running through all the options of what they have on hand to fix in a hurry, Cathy and I decide on baked ham, scalloped potatoes, green beans, and a sheet pan cheesecake for dessert. I ask her if she can do supper for us a few nights a week until the new dining hall is finished. She's happy for the business and tells me to say hi to Mr. Louis.

I decide it would probably be best to have a place to gather everyone together at once, so I contact a tent company and have them set up 2 big tents right in front of where the new dining hall will be. They even have the tables and chairs we need. I make arrangements to have the equipment stay here for a few weeks until the new hall is finished. Suddenly, I realize that I never ate lunch. I get really woozy and lay my forehead in my hand. Maia looks over and says, "Oh no, you don't!" and hands me a chocolate bar. I scarf it down and wait a few minutes. Then I stand to go see what I can find left of the pork chops from lunch.

The 3 guys are sitting down, still talking things over. Malcolm has a plate in front of him. I raise an eyebrow. I hear, "*Shit baby, I'm sorry.*" I wave a hand at him and find a pork chop waiting for me, still in the skillet. I just pick it up and eat it with my fingers. Then I set to tidying up the kitchen. *Who wants to come back to this mess tomorrow?*

When I'm finished and I am walking past the guys out the door, Mr. Louis stops me. "Ellie, were you able to get ahold of Cathy?"

I fill him in on all the arrangements I made, tell him Cathy said hello, and what Maia has scraped together for tomorrow's lunch. He seems pleased. "Thank you Ellie, please thank Maia as well."

Malcolm grabs my hand as I walk away, "Am I forgiven?" I kiss him on the forehead, "You were never in trouble."

###

I'm worn out, so I stop at Maia's desk to let her know I'm leaving early. I shut everything down and grab my purse. I walk down the good hallway back to the apartment. I stop, midstride in the hallway. *Wait a minute, Maia wasn't supposed to be back for a few days yet.* I hear in Maia's voice, *"Yeah, well, Bryan drove all night to get that stinky mess back home."*

I send back, "*Wasn't that illegal?*" She replies, "*Illegal as shit, and he did it anyway. Said no cop would stop him for long with that stinking mess.*" I'm laughing and walk up to the apartment door. Mr. Mittens is waiting to try and bust out again. I grab him and hold him up facing me. "What has gotten into you?"

I hear, "*No windows.*" In reply. Oh, I didn't think about that. I ruffle his fur backward, he always seemed to like that. "Let me ask Malky tomorrow if they would mind an office cat." Mr. Mittens rubs his face against mine and purrs. *Malky. Pfft.*

I run through the shower and lie down for a quick nap. But it wasn't quick, and I wake up; I dunno when. I throw on some exercise clothes and rush out the door, down the hallway, and through the office. When I arrive, the tents, tables, and chairs are all set up, and Cathy and her crew are serving food as fast as they can. It's like a picnic.

I turn my head when I hear an engine barrelling down the driveway. It's Bryan's truck without a trailer. He roars right up to the office and stomps inside. I decide I had better find out what's going on. After all, technically, I started this fiasco for him. I walk up to the front door, and there are dirty boot prints all the way through the office; some of it is squishy. Bryan exits Malcolm's office and throws up his hands.

"Where is thc jerk? I can't find him, and he has to help me." My face is blank because I have no idea what to do so I decide to try and summon Malcolm from wherever he is again. "*We need you here, Malky*." I get a reply, "*I'm on the way.*"

Malcolm walks in the front door and takes in the scene. I'm standing next to Bryan, who is positively covered in mud and you-know-what from head to toe. Malcolm's mouth drops open. *Better shut that quick if you don't want a full serving of the smell.* "Bryan, what happened?" Bryan slaps his dirty arm right over Malcolm's clean shoulder. "Let me show you…. FRIEND." Bryan motions to the truck for Malcolm to get in.

"*This can't be good.*" I send to him. "*It's not.*" I receive back from Malcolm, and he sends me a vision he saw in Bryan's mind: the poop trailer stuck in the garden area with its rear wheels buried in the mud. I decide I will work on the messy carpet after I eat. I feel for Malcolm and Bryan, but I have no idea how they will solve the trailer issue.

I get in line for dinner. Everybody seems very happy and content. I'm so thankful we were able to pull this off and give Chef, Grant, Cook, and Nova some extra help. Maybe with a few nights a week of dinner being solved, along with lunch every day, they can get some breathing room. When I get up to the line, I see a dark-haired, older woman with a limp. Her name tag says Cathy. As I'm going through the line, I introduce myself and thank her for coming through for us on such short notice. She smiles, "Mr. Louis and I have been friends for years. If he says he needs help, I will help." I smile back and thank her again. All the food looks delicious.

My plate is full and I'm walking down the aisles of chairs to find an empty spot. I find one in between some warehouse folks I have seen here and there. I introduce myself and ask if the seat is taken. My seat mates are Barney and Olivia. Barney is mid-30s, like myself, and has been working under Gerald for a few years now. Olivia is a more recent hire, just in the past year. I'd say she is in her early 20s, judging by her face. They both seem to like their jobs but complain about the brown uniforms. I tell them I will pass it on to see if we can get it changed. Olivia smiles wide and thanks me. After finishing their meals they both stay for a few minutes, making small talk. Eventually, they get pulled away by some co-workers to go get dessert. That leaves me sitting by myself. I finish my food and toss my paper plate into the big can. I see it is overflowing, so I tie it up and set it aside to pull up a fresh bag. Then I head back to the office to start on that carpet.

###

When I walk in, all the lights are on, and Nova is scrubbing at the spots on the carpet and cursing loudly. I'm pissed, "Nova! You are supposed to be taking the night off!" Nova stands and whirls around at me. "Yeah, Ellie, I'm supposed to be taking the night off, and I walk around through the office to find this mess. Nobody here to take responsibility and leaving the mess for me!"

Now my hackles are up. "Nova, now you just wait one minute. I wanted to eat before I tackled the mess. My intention was to return and get to scrubbing. I noticed the trash can was full and decided to reset it before coming in here." Nova is glaring at me with her arms crossed. I point outside, "Have a look! You can see the full bag sitting next to the can. I don't know where it goes, or I would have taken it there. I'm glad you're here so you can point me to the carpet machine. But I will not be called a liar like you just did. I know you can hear my mind; I'm not skilled enough to hide it."

Nova takes a deep breath and drops her arms to her sides. "Ellie, I'm tired." I'm still huffing mad, but look at her hard, "I know, Nova. That's why we are trying to help and do all this so you don't have to. Nobody wants you overworked and overtired." I point back to the tent. "Have you eaten yet?" She nods, "Yes, we were close to first in line. I don't know how to quit working. I knew there was going to be trash from tonight and I was fussing over the kitchen needing to be cleaned from lunch. But I went in there, and it was all done."

She puts her hand over her mouth and whispers, "Oh no. That was you, too." I nod and sigh. "Yes, Nova, I wanted you to truly have a night off. I don't know how you've been doing everything like you have." She steps forward and asks if she can hug me. I agree, and as soon as my arms are wrapped around her, she bursts into wracking sobs. *Whoa.*

Cook comes up behind me and pats me on the shoulder. "She finally break?" I nod and pass Nova off to him. He picks her up like a baby and carries her to the back of the office. I can only assume they are heading to the apartments, but I don't know. *SHIT!* I hear a chuckle, and in my mind, I hear, "*She keeps it in the storage closet in the front of the dining room.*" I go to the dining room, and inside the closet is the carpet machine. It's a huge green thing, very heavy looking.

Along the wall is a set of shelves, all neatly filled with cleaning products for various surfaces and labeled as such. *So organized.* I find the shelf for the carpet, and there's a spot cleaner, so I grab that. I also find a bottle of carpet shampoo. I locate the dispenser on the carpet machine and add the shampoo, so that's done. Then I back the big thing out of the closet. There's an entire shelf of paper towels, and I decide that would be best for cleaning up the initial part of the poo.

I push the big machine around the corner, carrying the supplies in my arm, and literally run right into Nathan. The carpet machine goes between his legs, and he tumbles forward into me. I've landed on one of the gooey foot prints. *Serves you right for suggesting the poo in the first place.* Nathan has landed on top of me; his leg is dangling over the top of the now sideways machine. He's groaning. The cleaning supplies flew everywhere.

Just then, a mud and poo-covered Malcolm walks in and sees quite a scene. The tangled bodies, the carpet machine, and scattered cleaning supplies have taken up the entire area. I take in a look of pure exhaustion coupled with frustration and now surprise that is on Malcolm's face. He leans down to me, "Are you hurt, baby?" I mentally check, "Everything seems ok. I'm afraid Nathan is hurt."

Malcolm reaches to try to lift Nathan off me. But Nathan is just dead weight. Malcolm leans down to whisper in his ear. "What is wrong, brother? What hurts?" I can hear in my mind, and I'm sure Malcolm does too, "Leg, think broken." We both look under the carpet machine to find one of Nathan's legs at an odd angle, partially under the machine. Malcolm lifts the machine up, and Nathan rolls onto his side.

Malcolm looks at me, and I interrupt him before he can speak, "Don't call Cook or Chef. See if the warehouse guys can help." Malcolm stands and turns to walk to the tent, but he is met at the door by 3 big guys. I don't know their names yet. One speaks for the group, "We've got this. Let us carry him." They lift Nathan, one at each end and one in the middle, and carry him to the infirmary.

I look up at Malcolm, "You stink." He looks down at me, "You don't know the half of it. Bryan almost broke up with me." Smiling, he reaches down for my hand and helps me up. I look up at him and put my hands flat on his chest, reaching for a kiss. He bends his head down for a light peck.

I smile, "Why don't you go clean up your stinky self, and I will finish this." I gesture to the carpet mess. He nods; he is really tired. As he walks away, he hollers, "Hurry every chance you get!" I giggle and turn back to my work. *Crap, I should have had him lift the carpet machine.* I pull and push and finally get it upright. None of the shampoo leaked, so that is good. I gather up all my supplies and set them on a spare desk. I start with the footprints in the foyer and work my way back. The spot cleaner works pretty well and gets most of the muck up with just paper towels. I run the hose from a spigot in the dining room closet, then connect it to the machine.

Just as I go to start up the machine, Mr. Louis walks in. "Ellie, what are you doing all that for?" He looks surprised. I smile at him and quickly relay the cow poo story. Mr. Louis roars with laughter and asks, "Do you know if the trailer got pulled out?" I reply, "I don't, but Malcolm is back and he's covered in it. I haven't seen or heard from Bryan, though." Then I stop his laughter while telling him of Nathan's broken leg. "Oh no. I will check on him. Thank you for handling all this, Ellie." He gestures to the floor and turns to go to the infirmary.

"Mr. Louis! Be careful of the hose!" I yell. I receive back, "*I saw it, Ellie, thank you.*" *Oh good.* I continue with the carpet machine, and there is only one little two-inch spot that I cannot get to come up, it's almost a green tint. *It must have been pure shit.* I sigh. I will ask Nova about it when I see her next. I clean up my mess, get the carpet machine stored, and roll up the hose. All items are back where I found them, and I shut off the lights. *I better go check on Nathan.*

***

CHAPTER 20:

I walk into the infirmary and see Kathleen. I smile wide and go up to her for a hug. "Hey, dear lady! I haven't seen you for a bit." She smiles and motions to Nathan, "He's buggered up for sure. That leg is going to need surgery." I stand back, "Who do we get for that?" Kathleen smiles, "No worries, Mr. Louis has already called a friend in St Louis. He's the top orthopedist in the country. He is flying up and will be here in a couple of hours. I've got plenty of pain meds already on board, so he is feeling nothing." She chuckles, "I imagine he is quite floaty at the moment."

I walk up to Nathan, smile at Mr. Louis in the side chair, and whisper in his ear. "You're forgiven. Thank you for bringing me here."

He reaches up and holds my hand for a moment, "Mom?" Mr. Louis laughs, "I didn't think of that, Ellie, until he just said it. You do have similar features to Belinda. She had darker hair, but your face shape and eyes are similar. Maybe that's why he was so taken with you from the beginning." I really don't know how to take any of that information. *I wonder if Malky thinks that?* I realize I'm quite tired, too. I look at Nathan, who is snoring and drooling on his pillow.

I look from Kathleen to Mr. Louis and say, "Folks, I think I've worn myself out today. I'm going to go crash." They both nod, and I walk the dark hallway from the infirmary to my apartment.

###

Malcolm is there, curled up on the couch with his cat. He's clean, but still looks exhausted. He holds up a hand. "I'll tell you the story later. I just want to snuggle and sleep. Is that okay?" I nod and head to the shower. When I get out, he is snuggled in the bed with his cat. I've never seen Mr. Mittens more content than when he is with Malcolm. *They are so in love.* Malcolm chuckles. "So am I baby, so am I." I blush and crawl in next to him. Both of us crash out.

The next morning, I wake up to an empty bed except for Mr. Mittens. He is tapping my nose with his paw to wake me up. I send with my mind, "*What is it?*" He sends back, "*No windows Mom.*" Crap, that's right, I was going to ask if he could come to the office. Malcolm calls from the kitchen; he must have heard us. "Yes, bring him. I will have Cook keep the door shut to the dining room." Mr. Mittens perks up and rushes out of the bedroom. I find him sitting at the front door like it's time to go right now. "Give me a little bit, buddy." I bend down to pat him on the head. Malcolm fixes some oatmeal for us and I get dressed for work. Malcolm walks to his apartment to get ready. We meet at the end of the hallway. I figure there's no sense in trying to carry Mr. Mittens that far and expect him to just follow us. He does. His little face is set and determined to be an office cat. *Put on a tie and give him a briefcase.*

Mr. Mittens and I drop Malcolm off at his office and walk to my desk. I point to Maia and tell him who she is. Maia exclaims, "THE KITTY! THE KITTY IS HERE!!!" Like she has wanted him there forever. I can hear Terrell's deep laugh from his office. Maia scoops Mr. Mittens up in her arms. I find an empty box and set it upside down on one of the reception chairs, right in front of the window, so Mr. Mittens can see everything that goes on. He instantly spots it and squirms out of Maia's grasp to go sit. I type up a sign for the dining room door. "Dining room open. Door closed to keep cat out. Don't let him in, no matter what he tells you!"

I walk over to said door and tape the sign to it. Hopefully that explains things. What happens though, is a barrage of people are so excited to see the cat; they trail through the office to stop and pet him before they go eat. It's like Mr. Mittens is the king and all his subjects have to pay homage before their meal. The little stinker just eats it up. I have a feeling he is going to be a permanent installation. I look over at him and he is beaming. I send to him, "*Nice upgrade from living under the shed out back.*" I get back, "*Sure is.*"

Maia fields calls and I work on the schedule all morning. I can see out front that 3 food trucks are pulling in and setting up near the big tent. I start to get excited about lunch, "Maia, what kinds of food today?" She turns to me, "It's going to change everyday. The BBQ guys said their business had been slow, so they will be here every weekday. The taco truck is coming Mondays, Wednesdays and Fridays; he said he can make more money in the store parking lots on Tuesdays and Thursdays. The Donut Lady is coming a couple days a week and agreed to pull in whenever it gets slow elsewhere. There's a baked ziti guy that is supposed to come, I haven't worked with him before and he said he would just show up when he could. And Mr. Barkley, the ice cream guy, is going to come every afternoon at 2 so he can catch us before he heads to the after-school crowd."

My eyes are wide, "That's a lot of options!" She interrupts me, "Oh wait, the little corn dog stand from the next town over is going to come on Saturdays to fill in for overtime folks." I raise my eyebrows, "Maia, right now that is all of us. Can he handle everybody?" She looks at me. "I couldn't think of anybody else. Do you have any ideas?" I shake my head. *Think think...* I look over at her, "I'm sure these food truck folks all know each other. Maybe they will have some ideas?" She nods, "I will ask them here in a few minutes. Monkey is out there helping them get hooked up to electric and water."

Malcolm walks up behind us. "Ladies, let's go check all this out." We all walk over to the food trucks. Malcolm goes to each truck in turn. He introduces himself and us, welcomes them and thanks them for coming. He goes on to say, "If you're slow at other times of the afternoon and evening, feel free to come back and hook back up. We can always use the extra grub." He stops at the BBQ Guys' set up and chats with them a bit longer. They show off their big smoker they brought along. It's a homemade job crafted out of an old anhydrous tank, so it's already on wheels. *How clever.* Malcolm points to it, "How many pounds of meat can you smoke at a time with that thing?" They go into detail about what and how much and Maia and I start to walk away.

Malcolm catches up to us. "Since they are going to leave their set up here round the clock, I've talked them into smoking a bunch of ribs for us one night a week. Could you coordinate that with Cathy?" I nod and go inside to make the call. We decide that Wednesday night would be rib night. Cathy is just tickled pink to collaborate with the BBQ Guys; she says they are one of her favorites. Maia is still worried about the weekends. I remind her that we still have 2 dining rooms and they are staffed. "Everybody knows the situation and don't want to put too much on Chef, Grant, Cook, and Nova. It will all work around who wants what every day and leftover BBQ is always better the next day anyway." She agrees and once it's lunchtime, we head out to try a little bit of everything. All 3 trucks have coordinated the buffet line, so it works just like last night, except there are different folks serving the food instead of Cathy's crew.

Mr. Louis walks by, his plate piled high with all the different options. "Maia, you've outdone yourself." Maia grins and Mr. Louis takes his plate inside. The trash cans are filling up again and I turn to Malcolm. "When I switch those out, where do the full bags go?" He rolls his eyes, "You are not going to take those bags. I will take them to the dumpsters up by the front of the drive." I must have looked puzzled. "There's a bank of dumpsters hidden behind the white fence at the entrance of the driveway." I apparently hadn't noticed it before. He rolls his eyes at me. I glare at him, "Stop that; there's punishment for contempt." He grins, "Oh really…" and waggles his eyebrows. Terrell is chuckling; Maia swats Malcolm,

"Ya'll are in public." I'm smiling too. It's great to see Maia and Terrell with smiles on their faces again.

The weekend and next week slide by. A sense of normalcy descends on us and the office gets into a groove. Mr. Mittens is really enjoying being the office cat. Terrell bought him a cat tree and it stands in the corner of his office. Mr. Mittens goes in there to nap most days, where it's quieter. Caleb is back to working all day each day and that takes some of the load off Malcolm. Nathan did great in surgery and has returned to the office on crutches. Nova hasn't appeared, so we all take turns vacuuming, dusting and keeping things tidy. Nobody has seen Bryan except Gerald and Cook. They have been meeting up in the night to ship out. *He must really be pissed.*

The large tents and tables outside are working out so well that the dining crews decide to just bring their food out in the morning and evenings now. The weather is crisp and fall-ish, but we haven't had much for rain, so as long as we wear jackets, it's quite pleasant. The new kitchen and dining hall building structure is completed and work is progressing on the inside now.

Malcolm hasn't stayed in my apartment for several days. We have been walking to work together and eating together every day. I find myself looking forward to seeing him and I feel that *zing!* every time he touches my hand. Heck, even when he uses his intuition to send a message to me. I have no idea what day it is at this point and I've decided to just run with it. I'm good with routine, so this feels more normal, even if there are no weekends right now.

After supper, Malcolm pulls me aside. "Can we walk and talk?" I nod and he takes my hand; we start walking toward the stream. "I want to let you know a few things that have been going on babe. I know you're not receiving messages yet." I nod. *I wonder when my brain will tune in?*

He twists my hand back and forth and smiles down at me. "It will. Let me fill you in." "Ok" is all I can say. "Dad has decided to start a private vitamin line. He has found a small natural supplements company to work with in Mississippi to package, market and distribute our mineral along with using the mineral in some of their own formulations. They are in a small town without a lot of employment options and it's a poor area. We are going to set up a new location there to make it simpler for the naturals company to access our product. There's a lady coming to interview tomorrow. She is going to help Maia a little with daily tasks and she will be the main coordinator for the Mississippi location since they are going to need a lot of product. Maia is going to be traveling more often to Oklahoma and I can't expect you to take over all of Maia's responsibilities on top of the additional scheduling you're going to have. The Mississippi lady's name is Ahbra." I look up at him, "That's a pretty name; what does it mean?" He smiles down at me, "let her tell it to you."

He interrupts my wondering. "There are 3 folks coming to interview for cleaning positions. One is for the new dining hall; his name is Redwood. Two ladies are interviewing to clean here in the office, warehouse, infirmary, apartments and original dining halls until they move across the drive. They are twins, Clara and Zara. We are also adding 5 warehouse folks and 5 to the harvesting crew, but you probably won't meet those folks for a little while." I squeeze his hand. "You've been busy." He chuckles, "No, Nathan and Dad have been busy searching for folks to add to the team. We got behind on hiring and it's put a toll on us all." We've walked the whole stream at this point and Malcolm turns around and starts to walk back. It's getting dark.

"And…" he says, "We are finishing up getting Oklahoma stocked and running. That means the extra overtime will come to an end and we will have weekends again." He sighs like he's nervous, "Maia and Terrell are going to take off this next weekend for a little getaway." I was wondering if the following weekend, maybe we could either go somewhere or atleast go out in the city?" My eyes widen, "Like a real date?" Malcolm chuckles and rubs his chin, "More like a mini vacation." I have to admit, my mind starts running at the thought of a break.

I try to act nonchalant, "Like, what are you thinking?" Malcolm goes off on a tangent about the different resort holdings they own. Finally, he turns back to me, "Would you want mountains or beaches?" I smile wide, "Well, I dunno. Can I think on it?" He grins, "Sure!" Using his mind, he sends me a vision of a bright white sand beach, clear blue water and salty sea air. Then he sends me one of driving up a mountain, it's almost vertical and I can feel the g-force pulling me backward. I see a lot of rocks. "What a choice!" I exclaim.

We walk back around to the office front door and I collect Mr. Mittens to go back to the apartment. Neither one of us wants to give up Malcolm tonight. I stop Malcolm, "How's the house coming?" He hangs his head. "It's at a complete standstill. The guys building the dining hall are the same ones who were working on the house." "Oh." is all I can say. Mr. Mittens trots behind us as we walk down the hallway. Malcolm stops at my door and kisses me tenderly. His voice is thick, "I will see you in the morning." I try to keep the tremble out of my voice, "Goodnight, baby." He turns to walk away and I watch that fine rear turn to enter his own apartment. Mr. Mittens meows at me; he wants his dinner now. I open the door and we step inside. *Ho hum, this separating every night is for the birds.*

###

I shuck off the day's clothes and sink into a hot bath. Then I wrap up in my robe, with a towel on my head and lay down in the bed for a few minutes to think. I don't get any thinking done because I fall asleep as soon as I settle in. Mr. Mittens wakes me up in the morning again. I look at him and send, "*What's wrong?*" He replies, "*Malky.*" I'm not sure what the issue is, but I get up, still in my bathrobe and go out the front door, Mr. Mittens trailing behind me.

I knock on Malcolm's door, but there is no answer. Mr. Mittens follows me in the door. I can hear some deep, almost sobbing sounds from the bedroom. We go in and I find Malcolm in bed, absolutely wrecked. He is sitting up in bed, head hanging low, his face red and swollen. There's a mountain of little wads of toilet paper he's used as tissues next to him. I fall into bed beside him and wrap him up in my arms. Mr. Mittens lays across Malcolm's legs.

At that moment, I can feel the fear rolling off of him. I have no idea what he is so afraid of but he grabs ahold of me and squeezes tight. I just hold him. I pat his hair and rub his back and just let him get it out. "Shhhh baby. You are ok. Whatever you're battling, we will do it together." I think I've said something good, but he just sobs harder. In my mind's eye, I can see flashes of faces of old folks that look like Mr. Louis; I see a face with features similar to mine with dark brown hair. *His mom.* I see him think about his dad and his brother. Bryan's face pops up and it lingers for a little while. There's the plans for the new warehouse being built in Oklahoma. A colorful label on a pill bottle flashes by. Then I can see the house, sitting all dark and lonely and unfinished. The view stays there for a while too. There's red cows being chased through a lot with cowboys on horses. *This man is just plain overloaded.*

Then a scene of him standing at the front of a church, looking back down the aisle at me, in a white dress. *Oh boy.* Then a picture of me holding a baby, then me walking a boy with brown hair along the creek. There's a little girl with pigtails having a tantrum at a table. I just keep patting him. *"Baby. We've got all the time in the world."* He finally starts to settle down.

I squirm out from under him and go to make some coffee. When I get back, he is sitting up and holding Mr. Mittens in his arms. He sets the cat down and I hand him the mug. He takes a sip like it's life-giving. In a hoarse voice, he asks, "How did you know to come in?" I hold his free hand, "Mr. Mittens woke me up." He smiles down at the cat and Mr. Mittens rolls over for belly rubs.

I look over at him and blurt out, "I think you should promote Terrell." He looks at me with wide eyes. I shrug. "Terrell has proven he is committed and you know he can handle juggling a bunch of different things. Your Dad is getting older, and you and Nathan are handling more and more. I think you should trust Terrell with some of these things." He gurgles a little and says, "We've never involved an outside person in major decisions for the company. It's always been Me, Nathan and Dad." I look at him, "You're wanting a change in your life. Making that change means that you're going to have to be available for your immediate family first. IF we get married and have babies, your priorities will have to change. Keep your pet projects and shift the rest over to Terrell so we can build a life together."

He is still sniffling and now his head is raised and he's looking at me with a dumbfounded expression. I can see the wheels turning and finally he says, "You're a pretty smart lady." I laugh, "I dunno about that, but I cannot bear to watch you go through torturing yourself again." He holds my hand and finishes his coffee. Mr. Mittens is still rolling around on the bed back and forth in the hopes somebody will rub his belly. I reach over and start patting him and run my fingers through his fur. "Don't start acting like a dog, Mr. Mittens. Keep the cat act up." Offended, Mr. Mittens slinks off the bed.

The caffeine is starting to kick in and Malcolm seems to come to life a little bit. He runs through the shower while I fix a little breakfast, just some bagels with cream cheese and fruit. While we are eating, I look over at him, deciding now is the right time to ask, "How did you and Bryan make out with the trailer full of shit?" Malcolm laughs, a big hearty laugh, "Yeah, that's quite a tale. You saw me leave with Bryan, but I ended up running back to the office alone. Bryan is still so angry with me. Not just for having to haul the poo, but also for getting stuck. I guess our proposed garden spot hangs wet." I groan.

He continues before I can comment, "I had to get the chainsaw out of the storage shed and cut a couple of the smaller trees down. Then we carried them over to the trailer and placed the logs in front of the wheels that weren't buried." He flashes me the same scene of the ass-end of the trailer buried to the top of the first set of wheels. I can hear Bryan yelling obscenities in the memory. He continues, "We had to get the shovels out of storage too. We dug and dug and dug until there was enough space in front of the buried wheels to wrap straps entirely around the trailer. You remember it was a rented trailer, right?" I nod. "I don't have a loader tractor purchased for the homestead yet. We had to dig everything by hand."

I slap my hand up onto my forehead and my mouth hangs open. Malcolm grimaces, "Yeah, Bryan yelled at me the entire time. It was horrible. I walked back over to the crane truck and thankfully, the keys were in it. I was able to lift the end of the trailer out of the mud using the crane. With it being level, Bryan was able to back his truck up and hook up to the trailer. I will never understand how he got the trailer unhooked from his rig in the first place." I'm trying to picture this in my mind. He sends me a scene of the crane truck holding up the the rear of the stuck and very muddy trailer, the wheels caked. Then I see Bryan's red truck backing under the hitch.

I say, "That makes sense. Was he able to pull it out then?" Malcolm shakes his head. "No, it was so muddy and mucky that the trailer just wanted to spin in place. The logs gave the wheels something to hold onto. I was still in the crane truck, holding the rear up so it wouldn't fall back down into the slop. Since the crane can articulate, I was able to back up at the same time as Bryan was pulling forward." He takes a deep breath. "Finally, the front wheels of the trailer caught on the logs and even though it sunk some, the logs gave the wheels enough traction to start pulling forward. I just kept backing up with the crane truck, continuing to hold the end of the trailer up. It was only about 10 feet but it felt like a mile at the snail's pace we were going. Soooo many things could have gone wrong."

He hangs his head, "After 10 feet, the ground was solid enough to hold the empty trailer. Bryan stopped long enough for me to unhook the crane and then he peeled out, with all the straps dragging behind him. I drove the crane truck back to where I found it and shut it down. I ran back to the office to get you." He smiles at me. "From what I understand, Bryan called Monkey and asked him to meet at the big truck wash up on the interstate. They used 2 of the pressure washers there to get the trailer clean before Bryan could return it. Monkey told me that Bryan wanted to get the thing returned right then. He hasn't been back to have Monkey service his truck at all. He has only been back to pick up loads and drop off empty trailers during the night. Cook says he has been taking food with him."

I'm shocked. "I had no idea Bryan had a temper like that." Malcolm sighs, "I haven't seen his temper flare like this since college." *What about his mind?* Malcolm replies to me out loud, "I can't hear it because he has been rolling in at night and we have been asleep." He pats my hand. "Cook says Bryan is still plenty pissed. I haven't called him yet to try and smooth things over. I'm waiting until Cook gives me the all-clear when he thinks Bryan might have softened a bit."

I wait a few moments before I ask, "What does the garden spot look like?" Malcolm rolls his eyes. "It was getting dark and I didn't look back. I haven't been back out there. Let's go for a ride later." I agree and he goes to get dressed. I leave Mr. Mittens with his Malky and walk back to my apartment to get dressed for the day. We are running a few minutes late.

CHAPTER 21:

The 3 of us walk into the office to find a whole pile of people waiting in the reception area. Maia looks stressed. Terrell is visiting with folks. Mr. Louis and Nathan are no where to be found. Mr. Mittens hides on his cat tree in Terrell's office; he wants none of this.

I use my mind, "*Gerald? Amira? Your people are here!*" Amira, the harvesting crew manager, arrives first. An average-height woman, she is still in her hardhat, coveralls and boots. She smiles at me and gives a thumbs up, "Thanks Ellie, I owe you one." I'm surprised by her British accent. Amira rounds up her 5 folks and heads back to the dining room to chat with them. Gerald arrives next and takes his 5 folks outside to visit and orient them to the operation. Maia looks over at me and whispers, "Thanks Ellie. I didn't know who went where and all their thoughts just had me jumbled." I chuckle, "I wouldn't have known if Malcolm hadn't told me about it last night."

I decide to step forward and introduce myself in turn to each of the 4 remaining folks. I start with a man sitting in the chair closest to the door. His legs are stretched out in front of him and he is sitting in the chair like it's way too short for him. *He has the longest legs I've ever seen!* I put out my hand to shake his and he shoots to his feet and pulls his ball cap off his head. "Ma'am." I look up and look up and look up some more. He smiles, "3 inches off of 7 foot ma'am." I smile back at him, taking in his red hair and red beard. He is dressed in jeans and a flannel shirt; his boots are well-worn. "I'm guessing you must be Mr. Redwood?" He chuckles, "My first name is Redwood, ma'am." He reaches for my hand and shakes it. I have 4 brothers, Oak, Chestnut, Ash and Hickory; we all look like this. He gestures to his body up and down. The only logical reply is, "Your Momma is a saint, 5 boys your size?" He laughs a big belly laugh, "She sure is ma'am; she is shorter than you and gave birth to us beasts. My Daddy was a lumberjack in his younger years; we lost him a year ago." I place my other hand over the one already in my hand. "Welcome Redwood; we are glad you're here." Malcolm and Cook come up behind me and introduce themselves. I move along to the next person.

As soon as I look at her I know, "You must be Ahbra!" She jumps up and gives me a big hug. "How did you know, lady?" I turn and wave Maia over so they can visit. Maia sits down next to Ahbra and strikes up a conversation. Ahbra is heavyset and rather tall. She's African American and has kept her hair natural, opting to keep it bundled behind a colorful head wrap with just the ends sticking out. She is dressed impeccably; sensible heels and a broomstick skirt with a big tribal pattern on it. She has on a deep brown button up satin shirt that perfectly coordinates with her skirt and is almost the same shade as her gorgeous skin. Her cheekbones are high and she is wearing black-rimmed glasses with plain gold long-drop chain earrings. *I'll always be the shortest lady in the office.* Ahbra turns back to me, reaches up and grabs my hand, "Tell me more about what you do, lady!"

I visit with Maia and Ahbra for another minute before walking to the corner. Two girls, they might be 21 by the look of them, are huddled in the corner whispering to each other. Sheets of long black hair cover their faces as they talk. I clear my throat and both look up at me doe-eyed. I can see Asian descent in their hair and coloring. Both have the creamiest clear skin I have ever seen. And they're identical. I smile wide at them. "Are you Clara and Zara?" Both giggle and reply, "Yes." At exactly the same time. I shake each of their hands in turn. "Welcome ladies, who is who?" There's another giggle in response. Their outfits are identical: jeans, flip-flops and white t-shirts, so that's no help. The one on the left says she is Clara, so I point at the other girl, "you're Zara then?" *This is going to be insane keeping this straight.*

I hear Nova coming up behind me and her loud ramblings about being the only one that cares, etc… etc… I turn in time to wrap that woman up in a hug as big as I've got. When I let go, she backs away a step and grins at me. "Missed me, huh?" I smile wide, "Sure did lady. Looks like you've got your hands full." I motion to the 2 girls and turn to them. "Clara and Zara, this is Nova; she will keep you in line." Both the girls giggle again and stand. Nova leads them away to go talk.

I slap my hands on my hips and go back to my desk. I've run out of people to talk to. Maia stands and brings Ahbra over. "Ellie, could you take Ahbra back to the conference room? I need a minute." Understanding she is needing a trip to the restroom, I lead Ahbra back to the conference room and regale her with my tale involving the mini fridge. She has a big laugh to go with her personality. *She's gonna liven this place up.* Mr. Louis comes in behind us and retells the story of rescuing me from the angry mini fridge.

I leave the 2 of them chuckling and stop at Malcolm's office, but he is gone. He left his phone, and it's ringing. I let it go to voicemail and it rings again. I pick the phone up to take it to Caleb, he is missing and then rush over to Terrell's office; he is missing. I know Nathan is going to be talking to Ahbra, so I decide to answer it. It's a woman rattling off the head as fast as she can in Italian. I hold the phone away from my ear to look at the screen to see what the number is saved under. It is just a number; Malcolm hasn't saved the number in his phone before. The woman is still rambling and I go back to the call log and see that she has called 15 times already today but hasn't called previously. *Already.* I hold the phone back up to my ear, and mumble, "I'm sorry ma'am, he will have to call you back." Then I push the end call button. I lay the phone back down and walk back to my desk. *Whatever she was saying, she was pissed with a capital P.*

I decide to turn around and check on Mr. Mittens quick. In that amount of time, Malcolm has walked through the front door with his lips pursed and a scowl on his face. I almost run into him. He scowls down at me and keeps walking. *Well... that was nice.* I keep walking to check on Mr. Mittens and find him curled up with his paw over his face. All the chaos hasn't bothered him a bit. I get settled back at my desk and start working in Volley.

Malcolm goes stomping out the front door and once again, doesn't say anything to me. This happens 2 more times. I look at him each time and he very pointedly doesn't look at me. *What's up with all that?* Maia comes back to her desk and she has brought me a latte. "Oh you sweet woman," I say. She grins and sits back down at her desk. Malcolm stomps by again. Maia whispers over at me, "What's up with that?" I shrug, mumble, "I dunno." And keep working.

###

Lunchtime rolls around. No Malcolm. *Whatever.* Terrell and Caleb haven't come back to the office either. Maia and I head out to the tent for lunch. It's a light drizzle today, so we hurry through the buffet and get back to our desks. The rest of the day is pretty low-key. I'm thankful it's Friday. A whole weekend and I have no idea what I'm going to do with it! I remember that Maia and Terrell are going away for the weekend and turn to visit with her and realize that she is gone now too. *All right, I've had enough.* Everybody is missing. *I'm left in here by myself and I have no idea what's going on.* I get up and go look for everyone.

I walk past the conference room and see Nathan and Ahbra visiting. I can overhear him, "You will need to be prepared to immediately move onto the premises here." Then I realize that I have no idea where anyone else really lives, well, besides Malcolm, Nathan and Terrell anyway. There has to be a larger amount of housing somewhere for all of us. *Why hasn't anyone talked about it?*

I turn around and head back past the dining room and see Cook and Nova preparing for dinner. I don't see Clara and Zara with them. I walk out into the warehouse; nobody here. *What is going on?*

Something tells me to head up to the staging area. I walk up the ramp and see absolute bedlam. There is white mineral in an avalanche coming from the main doorway of the chamber and spilling into the staging area. There is a mound of it piled in front of the infirmary door. There's people everywhere, trying to scoop up the mineral and get it sent on the ATV train into the bagging area. The ATVs can't keep up. It's spilling out of the door and coming faster than everyone can get it scooped. I stop at the staging area, switch out my shoes for boots and wade into the mass.

In front of me are Malcolm, Terrell and Caleb, all staring at the ceiling. It doesn't look like the ceiling caved in, but more like it's snowing. And the mineral is snowing down so hard, an avalanche is coming out the front door of the harvesting chamber. All of the warehouse folks are in boots and hardhats scooping; all 10 of the new hires are in boots and hard hats scooping. Even Clara and Zara are trying their best to help but are almost having a snowball fight with it at the same time. *Those two are going to be a handful.*

I walk up behind Malcolm and tap him on the shoulder. He whips around and glares down at me like I've done something wrong. I turn to Terrell instead. "What is going on?" Terrell yells back over the noise, "We aren't really sure yet!" He turns away and I tap him on the shoulder again, "Can I help anywhere?" Terrell shakes his head no, so I go back to the staging area, take off my boots and hat and head down to Chef's dining room.

I find Maia inside talking to Chef and Grant. I am apparently interrupting an intense discussion. "Hi guys." I look at the group as a whole. Chef stands and pulls out a chair for me. I look up at him, "Thank you Chef. Do you know what is going on out there?" Maia interjects, "That's a huge mess is what that is." I look at her in puzzlement. "Maia, I'm afraid you're going to have to fill me in from the beginning."

She sighs, "I'm sorry Ellie, I forget that you haven't been here forever. Do you remember when you took your tour of the facility?" I nod. She continues, "Do you remember the big drill rigs positioned around the harvesting chamber?" I nod again. "Those are fake." My mouth forms a little o. I look at her, puzzled, and she starts in, "Those are just there for the government inspectors to be satisfied when they show up to look at a *quarry* or a *mining* operation." She put quarry and mining both in air quotes.

I am so confused. "Ok?" She shakes her head, "No, I don't think you're getting it. The mineral self-refreshes each day. It literally forms on the inside of the walls and falls down, like the wall is creating it. The harvesting crew just comes in and scoops it up each day, sends it to the bagging area and then to the warehouse." She looks deep into my eyes to see if I'm tuning in. "You're understanding what I mean, right?" I nod, "We never run out; God provides the amount we need everyday." She smiles, "Exactly."

Finally, it dawns on me, "It's like manna from Heaven."

Author's Note: Shattered Intuition is a work of fiction. Any similarity between the story line and actual events, people, places, and things is entirely coincidental.

Made in United States
Troutdale, OR
11/03/2026

56845387R00130